Philosophy Beyond the Physical

A Guide to Metaphysics

Trenton H. P. Williams

Philosophy Beyond the Physical

TABLE OF CONTENTS

Philosophy Beyond the Physical

Introduction to Metaphysics

Metaphysics is a branch of philosophy that dives deep into the nature of reality, existence, and the fundamental principles that underlie all things. It deals with questions that go beyond the physical and measurable, seeking to understand the essence of being and the fabric of reality itself. From exploring the nature of reality to questioning the structure of time and space, metaphysics challenges our perceptions, inviting us to contemplate the mysteries of existence and our role within it.

What is Metaphysics?

The term "metaphysics" originates from the Greek word *meta*, meaning "beyond," and *physika*, referring to the physical or natural world. In essence, metaphysics is the study of what lies "beyond the physical." While science investigates the material aspects of reality—things that can be observed, measured, and tested—metaphysics goes a step further. It seeks to understand the underlying principles and the unseen forces that shape reality, often posing questions that cannot be directly answered by empirical evidence alone.

At its core, metaphysics investigates questions that are both universal and profound:

- **What is the nature of existence?**
- **What does it mean to "be"?**
- **What is reality, and how is it structured?**
- **What is the nature of time, space, and causality?**
- **Is there a purpose behind existence?**

To make sense of these abstract inquiries, metaphysics is often divided into several branches:

1. **Ontology:** the study of being, existence, and the nature of reality.
2. **Cosmology:** the exploration of the origin, structure, and purpose of the universe.
3. **Epistemology (when intersecting with metaphysics):** the study of how we know what we know, especially concerning what can be known beyond the physical senses.
4. **Theology:** as it relates to metaphysical questions about divinity, the existence of God, and the role of the divine.

Metaphysics is not merely theoretical but forms a lens through which many interpret the world, think about the purpose of life, and seek answers to questions that science may never fully answer. It is a quest to understand the "why" behind existence, encouraging deep reflection on life and our role within the cosmos.

The History and Evolution of Metaphysical Thought

Metaphysical inquiry has a rich history stretching back thousands of years, evolving alongside human thought and civilization. Early philosophers in Ancient Greece, such as **Thales, Pythagoras, and Heraclitus**, began questioning the nature of reality, formulating ideas that would lay the foundation of metaphysical thought. However, it was **Aristotle** who would formalise the study of metaphysics. In his works, particularly in a collection of writings later grouped as *Metaphysics*, Aristotle delved into questions about existence, substance, form, and causality. His ideas about "first causes" and "the unmoved mover" (a concept that can be interpreted as a prime cause or God) influenced thinkers for centuries.

During the Middle Ages, metaphysics became deeply intertwined with theology. **Thomas Aquinas**, for example, combined Aristotelian metaphysics with Christian doctrine, seeking to reconcile faith with reason. The concept of an ultimate, divine cause behind existence became central to metaphysical discourse.

With the **Renaissance** and the advent of **modern philosophy**, metaphysics encountered new challenges. **René Descartes** introduced a dualistic view of existence, separating mind and matter, which became a central debate in metaphysics. Descartes' famous assertion, "Cogito, ergo sum" (I think, therefore I am), brought the focus to human consciousness and self-awareness as the starting point for understanding existence.

Immanuel Kant, another pivotal figure in metaphysics, argued that human knowledge is limited by our perceptions. He introduced the idea of phenomena (what we experience) versus noumena (things as they are independently of our perceptions), challenging the notion that we can fully understand reality. Kant's insights added a new layer of complexity to metaphysics, emphasising the subjective nature of knowledge.

In the **20th century**, metaphysics continued to evolve, with thinkers like **Martin Heidegger** and **Jean-Paul Sartre** exploring questions about human existence, freedom, and meaning. Meanwhile, advancements in **quantum physics** introduced concepts that blurred the lines between science and metaphysics, especially in areas like the nature of particles, the uncertainty principle, and the role of the observer in defining reality.

Today, metaphysical questions remain relevant, with intersections in fields as diverse as philosophy, psychology, spirituality, and theoretical physics. Metaphysical inquiry has expanded beyond academia, entering popular discourse as people continue to seek meaning, purpose, and understanding of their place within the universe.

Why Study Metaphysics?

Studying metaphysics may seem abstract or esoteric, but it offers profound benefits that go beyond theoretical knowledge. Here are some of the compelling reasons why studying metaphysics is valuable:

1. **Deepening Understanding of Existence:** Metaphysics tackles some of life's biggest questions, encouraging us to explore what it means to be, to exist, and to experience. In grappling with concepts of reality, space, time, and being, we gain a richer, more nuanced understanding of life itself.

2. **Fostering Critical Thinking and Open-Mindedness:** Metaphysical questions are often open-ended, with multiple perspectives offering valuable insights. Engaging with these ideas fosters critical thinking, urging individuals to consider different viewpoints and to recognize that there may be more than one "truth" about reality.

3. **Connecting Science and Spirituality:** Metaphysics bridges science and

spirituality, allowing for exploration of questions that science alone may not answer. Aspects of metaphysics, such as the nature of consciousness or the origins of the universe, are at the forefront of scientific investigation but also deeply resonate with spiritual traditions.

4. **Enhancing Self-Awareness and Personal Growth:** At its core, metaphysics encourages introspection and self-reflection. By questioning the nature of our own existence and the workings of consciousness, we develop greater self-awareness, potentially leading to deeper fulfilment and personal growth.

5. **Seeking Purpose and Meaning in Life:** Many turn to metaphysics as a means of finding purpose. By questioning why we exist, what reality is, and whether there is an overarching design to life, metaphysics provides a framework for individuals to explore and define their own sense of purpose and meaning.

6. **Understanding Reality Beyond the Physical:** In studying metaphysics, we look beyond material existence to consider intangible elements such as energy, spirit, and the interconnectedness of all things. This exploration may lead to a richer appreciation for the invisible forces that shape our lives, encouraging us to live more consciously and connectedly.

7. **Exploring Consciousness and Human Potential:** Metaphysics opens doors to understanding consciousness and the potential of the human mind. By exploring ideas about mental energy, intuition, and non-physical realities, individuals can develop tools for personal empowerment, creativity, and transformation.

In conclusion, metaphysics is more than a branch of philosophy; it is a journey into the fundamental nature of everything that exists. By studying metaphysics, we gain insights not only into the universe but also into ourselves, deepening our understanding of what it means to be human. Whether one approaches it as a philosophical discipline, a spiritual pursuit, or a means of personal exploration, metaphysics provides an invaluable framework for pondering the mysteries of existence.

4o

You said:

Write extensively on Reality and Existence Understanding Reality: The Nature of Existence The Material and Non-Material Worlds Appearance vs. Reality: What Do We Really Know?

ChatGPT said:

ChatGPT

Reality and Existence

Reality and existence are central topics in metaphysics, where philosophers and thinkers examine the foundations of what is real, what it means to exist, and how we come to understand the nature of our reality. These questions touch on the boundaries between perception, material and immaterial realms, and the deep mysteries of consciousness. Exploring reality and existence involves questioning what we often take for granted and challenging the assumptions that underlie our understanding of the world.

Understanding Reality: The Nature of Existence

To explore the nature of existence is to ask, "What does it mean for something to exist?" Existence is often considered the most basic condition for anything, but its nature is complex and multi-dimensional. At a fundamental level, "existence" refers to the state of "being" rather than "non-being." For instance, a person, an object, or even an idea can be said to exist, but each exists in different ways. This raises the question of whether different kinds of existence might require different definitions or explanations.

Philosophers have long debated whether existence is something intrinsic to things themselves or something attributed to them by observers. **Plato** argued that reality consists of ideal forms that exist beyond our physical world; these forms represent the truest essence of things, whereas the objects we perceive are mere shadows of these ideals. **Aristotle**, on the other hand, believed that reality is rooted in the tangible, material world, where each entity has both a form and a substance, existing as unique individuals within the universe.

Existentialists like **Jean-Paul Sartre** and **Martin Heidegger** brought another perspective, emphasising that existence precedes essence. They proposed that individuals must create their own meaning, as existence itself has no predetermined purpose. In their view, reality is something each person must grapple with personally, highlighting the importance of individual consciousness in determining the nature of one's reality.

This search for understanding continues today, with modern theories in science and philosophy delving into questions of **consciousness, self-awareness**, and even **quantum mechanics**. These fields explore how the mind perceives reality and whether an observer can influence the state of existence itself. Metaphysical questions about existence therefore reach beyond physical explanations, inviting us to consider subjective experiences, consciousness, and even possible alternate dimensions as essential parts of reality.

The Material and Non-Material Worlds

One of the foundational inquiries in metaphysics concerns the relationship between the **material** and **non-material** (or immaterial) aspects of existence. Material reality includes everything that can be observed, measured, and interacted with—such as physical objects, energy, and forces. The non-material realm, however, encompasses abstract concepts, consciousness, emotions, and possibly spiritual elements, which are not easily quantifiable or observable.

1. **Material Reality:** In materialism, a major school of thought, the physical world is considered the only true reality. Materialists assert that everything can be explained by physical processes, with consciousness and thoughts emerging from brain activity. This perspective has driven scientific inquiry for centuries, forming the basis of disciplines like biology, chemistry, and physics, which seek to explain the world by understanding matter and energy.

2. **Non-Material (Immaterial) Reality:** The non-material realm challenges materialistic explanations by positing that there are aspects of existence that are not rooted in physical matter. Concepts like **consciousness, emotions, love, beauty, and spiritual experiences** don't have a physical form yet are widely acknowledged to be real in some sense. In

many metaphysical frameworks, these aspects are seen as fundamental to existence and may even be considered as forming their own plane of reality. Idealism, a metaphysical theory advanced by philosophers like **George Berkeley** and **Immanuel Kant**, posits that reality is primarily mental or spiritual in nature. According to this view, the material world may merely be a manifestation of the mind, with consciousness as the core essence of reality.

3. **Dualism:** Dualism, most famously associated with **René Descartes**, argues for a dual nature of existence, comprising both mind (non-material) and matter (material). In Descartes' view, the mind and body are separate entities that interact but are fundamentally distinct. Dualism has influenced metaphysical discussions about the nature of consciousness, where questions arise about how an immaterial mind can influence a physical body.

Today, these inquiries extend into fields such as **quantum mechanics and theoretical physics,** where scientists question the boundary between the observer and the observed, and explore phenomena that suggest the material and non-material worlds may be more intertwined than previously thought. The question remains: is the mind just a product of material processes, or does it reflect a deeper, non-material reality?

Appearance vs. Reality: What Do We Really Know?

The distinction between appearance and reality is another core area in metaphysics, probing how accurately we can perceive the world and whether our senses deceive us. **Appearance** refers to how things seem to us—what we perceive with our senses—while **reality** is the objective state of things as they actually are, independent of perception.

1. **The Limits of Perception:** Human perception is limited by our senses and by cognitive filters that shape how we experience the world. For instance, our eyes can only detect a small spectrum of light, meaning that much of reality is invisible to us. Similarly, our understanding is shaped by language, culture, and prior experiences, creating a version of reality that is highly subjective. **Philosophers like Immanuel Kant** suggested that our experiences are inherently mediated by these subjective filters, and that the true nature of reality (what he called the "noumenon") remains ultimately unknowable.

2. **Plato's Allegory of the Cave:** Plato's famous allegory of the cave provides a powerful metaphor for the relationship between appearance and reality. In the allegory, prisoners are chained in a cave, only able to see shadows projected on a wall. The shadows represent appearances, while the true objects outside the cave

represent reality. According to Plato, most people live in a state of ignorance, mistaking appearances for reality, and only through philosophical inquiry can one "exit the cave" and gain insight into the true nature of existence.

3. **Modern Science and Reality:** Science continually reveals that the universe is more complex and strange than it appears. Quantum mechanics, for example, challenges our understanding of reality at the subatomic level, showing that particles can exist in multiple states simultaneously and only "solidify" into a definite state when observed. This phenomenon, known as the **observer effect**, suggests that reality might not be fully objective and may instead depend, in part, on observation and consciousness. Such findings blur the line between appearance and reality, raising questions about whether the universe we observe is truly "there" or a construct shaped by our minds.

4. **The Nature of Illusion and Deception:** Another aspect of appearance versus reality is the potential for illusions, biases, and even outright deception. Optical illusions, for instance, demonstrate that our senses can easily be tricked. Similarly, cognitive biases and emotional states can alter our perception of events and people, creating a version of reality that might be deeply skewed. This has led philosophers to ask: if our senses and minds are prone to error, can we ever fully trust our perception of reality?

5. **Solipsism and Scepticism:** Solipsism is the philosophical idea that only one's mind is certain to exist. From a solipsistic perspective, everything outside one's consciousness may only be an illusion or projection. This extreme scepticism challenges the assumption that external reality exists independently of our perceptions. While few philosophers fully endorse solipsism, it serves as a provocative reminder of how limited and

potentially unreliable our knowledge of the external world might be.

The Nature of Being

The concept of "being" is foundational to metaphysics, encompassing the essential nature of existence, the qualities that define entities, and the nature of consciousness. To explore the nature of being is to ask profound questions about what it means to "be," both for individual entities and for existence as a whole. This topic looks into the study of **ontology**—the branch of philosophy that examines the nature of existence—and the exploration of **consciousness and self-awareness**, which introduces questions about the self, identity, and subjective experience.

What Does it Mean to "Be"?

The question "What does it mean to 'be'?" is deceptively simple, yet it requires a deep investigation into the qualities that make something exist. "Being" refers to the existence of something, whether it is an object, a person, a concept, or an idea. However, to "be" is not only to exist but to possess an inherent nature or essence that makes something what it fundamentally is.

The question of being is central to human understanding of reality, as everything we experience presumes the existence of entities with distinct forms, identities, and natures. In philosophy, this inquiry often distinguishes between different modes or types of being, such as:

1. **Physical Being:** This refers to things that exist in a material, tangible way, like rocks, trees, planets, and animals. Physical beings have a form, location, and can be

measured, observed, and analysed through the sciences.

2. **Abstract or Conceptual Being:** Ideas, numbers, and concepts also exist in a certain sense, although they lack physical substance. They are understood to exist within the mind, or as part of a larger intellectual framework, yet they impact how people think, reason, and organise knowledge.

3. **Personal or Conscious Being:** Self-aware beings, like humans, have a unique experience of existence. This kind of being involves self-reflection, perception, emotions, and a sense of individuality. Philosophers often see this kind of being as more complex and multifaceted, incorporating subjective experience and the capacity for introspection.

4. **Potential Being:** Potentiality, a concept from Aristotle, refers to the possibility of being something else. For example, a seed has the potential to become a tree, and a person has the potential to gain knowledge

or skills. This notion of potentiality plays a key role in understanding the dynamic nature of existence, suggesting that "being" is not always a static state but can involve growth, change, and development.

The essence of being, or "being-ness," is a quality shared by all entities. To say something "is" means that it has achieved a certain presence in reality, regardless of its form or the manner of its existence. **Martin Heidegger**, a prominent philosopher, explored this concept in depth, especially in his work *Being and Time*. He argued that most people take the nature of being for granted and that genuine understanding requires an examination of "being" itself, not just the particular beings within reality. This leads to questions of **purpose**, **essence**, and the **significance** of existence, which continue to be central to metaphysical inquiry.

Understanding Ontology

Ontology is the branch of metaphysics dedicated to understanding the nature and structure of existence. It addresses the most fundamental questions of being by examining what entities exist, how they can be categorised, and what the relationships are between different types of beings.

1. **Ontological Categories:** Ontology attempts to categorise existence by creating distinctions between different types of entities. For example, it differentiates between physical objects, such as a chair, and abstract concepts, such as justice or beauty. Ontologists also distinguish between individuals (particular beings like a specific cat or a particular book) and universals (qualities or properties like "cat-ness" or "book-ness" that can be shared by multiple things).

2. **The Nature of Substance:** A central question in ontology concerns what constitutes the "substance" of an entity. For example, Aristotle defined substance

as that which exists independently and can contain properties. The substance of a tree, for instance, is the core essence of "tree-ness," which distinguishes it from other forms of life. In contrast, properties such as colour, height, and location are considered **accidental attributes** of that substance, as they can change without altering the essence of the tree itself.

3. **Ontological Dualism and Monism:** Ontologists also consider whether there are multiple types of substance (dualism) or a single, unified substance underlying all of reality (monism). Dualism, a view advanced by **René Descartes**, posits that reality consists of two separate substances: mind and matter. Monism, on the other hand, suggests that all existence is derived from a single source or substance. Philosophers like **Spinoza** argued that everything is a manifestation of a single substance, often identified as God or nature.

4. **Existential Ontology and Human Being:**
   Ontology isn't limited to the
   categorization of objects; it also considers
   the unique nature of human existence.
   Existential ontology examines the special
   quality of human life, particularly the way
   individuals perceive themselves and their
   surroundings. Unlike inanimate objects,
   humans can reflect on their own existence,
   an ability that introduces concepts like
   **authenticity, freedom, and
   responsibility**. For existentialist
   philosophers like Heidegger and Sartre,
   ontology addresses not only the essence of
   things but the profound and often
   challenging nature of human existence.

Ontology thus provides a framework for
understanding reality on a foundational level,
addressing questions about what truly exists, the
structure of reality, and how different forms of
being relate to one another. It serves as a lens
through which we can interpret the complexities
of existence and our place within the world.

Exploring Consciousness and Self-Awareness

Consciousness and self-awareness are among the most profound and mysterious aspects of being, raising questions about the nature of the mind, the self, and subjective experience. While **consciousness** refers to the ability to experience awareness, **self-awareness** is the recognition of oneself as a distinct, individual entity with thoughts, emotions, and intentions.

1. **The Mystery of Consciousness:**
   Consciousness is the state of being aware of one's surroundings, thoughts, and feelings. It is a continuous, subjective experience that shapes how we interpret reality. Despite centuries of inquiry, the nature of consciousness remains elusive. **Philosophers like David Chalmers** refer to this as the "hard problem of consciousness," asking how and why subjective experiences arise from physical processes within the brain.

Consciousness is also tied to the concept of **qualia**, which are the individual instances of subjective, conscious experience—like the redness of a rose or the taste of chocolate. These qualia are unique to each person, and their subjective nature makes them difficult to quantify or explain purely through physical science.

2. **The Self and Identity:** Self-awareness involves not only consciousness but a sense of self, or personal identity. This includes an understanding of oneself as a consistent, autonomous being with a personal history and a unique perspective. **John Locke**, an early modern philosopher, argued that personal identity is tied to memory—our continuity of experience over time. In other words, you are the same person today that you were yesterday because you remember your past experiences as your own.
The concept of the self also raises questions about the boundaries of identity. Is the self simply the brain, or does it

extend to the body, environment, or even relationships? Some philosophers, such as **David Hume**, argued that there is no true "self" beyond a collection of perceptions and experiences, challenging the idea of a permanent, unchanging identity.

3. **Mind-Body Relationship:** The relationship between mind and body is another area of interest in metaphysics. Dualism proposes that the mind and body are separate, with the mind as a non-physical entity capable of consciousness and thought, while the body is material. This view raises questions about how an immaterial mind can interact with a physical body, an issue known as the **mind-body problem**. Conversely, **materialism** argues that consciousness is a byproduct of physical processes in the brain. This perspective suggests that self-awareness and consciousness are rooted in biological functions and that our sense of individuality is a result of brain activity.

However, this view does not easily explain the subjective nature of conscious experience.

4. **The Role of Self-Awareness in Existence:** Self-awareness is considered by some philosophers to be what sets humans apart from other beings, allowing for complex thought, moral consideration, and the search for purpose. The awareness of one's existence, mortality, and limitations introduces the potential for existential angst and reflection on life's meaning. **Existentialist thinkers**, like Sartre and Kierkegaard, argued that self-awareness compels individuals to confront their freedom and responsibility in shaping their lives, emphasising the uniqueness of human existence.

5. **Collective Consciousness:** Some theories propose that consciousness might not be entirely individual, suggesting that a form of **collective consciousness** might exist, connecting individuals in subtle ways. The idea is found in spiritual and philosophical

traditions, such as in **Jungian psychology**, where Carl Jung proposed the concept of the **collective unconscious**—a set of shared symbols, archetypes, and memories that influence all human minds. Such theories blur the line between individual and collective experience, suggesting that being is both a personal and shared phenomenon.

Space, Time, and the Universe

Exploring space, time, and the universe in metaphysics brings us to questions about the fabric of reality itself, our perception of existence, and our place within the cosmos. These concepts are not only central to physics but are also deeply philosophical. They shape how we understand everything from movement and change to existence and the nature of the cosmos itself.

The Nature of Space: Is It Real or an Illusion?

Space is the arena within which all physical objects exist and interact, but its true nature remains a topic of debate. Is space something tangible, with a real existence independent of objects? Or is it simply a construct of human perception—an illusion that helps us organise our experiences?

1. **Absolute vs. Relational Views of Space:** In metaphysics, the nature of space has often been debated as either **absolute** or **relational**. The absolute view, championed by Sir **Isaac Newton**, posits that space is an actual, infinite "container" that exists independently of the objects within it. According to this view, space would continue to exist even if there were no objects to fill it, making it a constant and unchanging entity.

   In contrast, the relational view, supported by **Gottfried Wilhelm Leibniz**, argues that space is nothing more than the relationships between objects. If there were no objects to relate to one another, space would have no meaning or existence. According to this view, space is not a "thing" but rather a way of describing the distances and orientations of objects relative to one another.

2. **Space as a Fabric or Structure:** Modern physics, particularly through the theory of **general relativity**, has introduced the idea of space as a kind of "fabric" that can be curved, stretched, or warped by the presence of massive objects. **Albert Einstein** showed that space (and time) are part of a unified "space-time" continuum, which can be influenced by gravity. This theory challenges the notion of space as empty or purely relational, as it implies that space has physical properties that respond to mass and energy. In this sense, space can be seen as an almost tangible structure that is integral to the existence of objects.

3. **Space as an Illusion or Mental Construct:** From a metaphysical perspective, some argue that space may

not exist outside human perception. **Immanuel Kant** proposed that space (and time) are forms of human intuition, mental constructs that allow us to organise sensory experiences in a coherent way. According to Kant, we do not experience "space" as it truly is, but rather as it appears to us through our mental faculties. This perspective suggests that space might be more of an illusion created by the mind than an actual, external reality.

4. **Quantum Physics and Nonlocality:** The strange implications of quantum physics have further complicated our understanding of space. **Quantum entanglement**, for instance, implies that particles can influence each other instantaneously, regardless of the distance between them. This phenomenon, known as "nonlocality," challenges the traditional concept of space as a separator between objects and suggests that space may be interconnected in ways we don't yet fully understand. Some philosophers and

physicists speculate that space, at a
fundamental level, may be emergent from
deeper, underlying principles rather than
an independent "container."

The question of whether space is real or an
illusion remains unresolved, but it opens up
possibilities for understanding reality beyond
traditional three-dimensional thinking. By
examining space metaphysically, we consider
whether the distances and separations we
experience are intrinsic to reality or simply
aspects of human cognition.

Time: A Perception or an Absolute?

Time, like space, is fundamental to our
experience of reality. It governs the sequence of
events and our sense of past, present, and future.
Yet, time's true nature is one of the greatest
mysteries in metaphysics. Is time an absolute
flow independent of human experience, or is it a
product of consciousness, shaped by the mind?

1. **Time as Absolute:** For centuries, philosophers and scientists alike have considered time as an absolute entity. **Newton** viewed time as a continuous and unchanging flow, ticking forward uniformly, regardless of the events within it. This view supports the idea of a universal, objective timeline that applies to all beings and events equally. This concept of time aligns with common-sense experience, as we tend to think of time as a constant backdrop against which everything happens.

2. **Relativity and the Fabric of Space-Time:** Einstein's theory of relativity transformed our understanding of time, demonstrating that it is not a separate entity but interwoven with space in a four-dimensional continuum called **space-time**. Time, according to relativity, is not constant; it can speed up or slow down relative to the position and velocity of objects in space. This leads to phenomena such as time dilation, where

time moves more slowly for objects in strong gravitational fields or at high velocities. This concept suggests that time is not absolute but is affected by the fabric of the universe itself.

3. **The Psychological and Subjective Experience of Time:** Philosophers also consider the possibility that time may be primarily a mental construct, existing only as a framework for human perception. **Henri Bergson**, a French philosopher, argued that time is deeply tied to human consciousness and that the lived experience of time—its flow, its rhythm—is distinct from objective, measurable time. To Bergson, time is not a series of discrete moments but a continuous, flowing "duration" that shapes human experience in profound ways. This view highlights the distinction between "objective" time (which can be measured by clocks) and "subjective" time, which varies based on individual perception and emotion.

4. **The Concept of "Now" and the Illusion of Flowing Time:** Some philosophers, including **J.M.E. McTaggart**, argue that time is an illusion, challenging the reality of past, present, and future. McTaggart proposed that events are not objectively "in time" but that time itself is an illusion created by the human mind. From this perspective, the present moment or "now" does not move; rather, it is the mind that constructs a sense of flowing time. This aligns with certain interpretations in physics, such as the **block universe theory**, which suggests that past, present, and future all coexist in a "block" of space-time, with no real "movement" through time.

5. **Quantum Mechanics and the Flow of Time:** Quantum mechanics adds further complexity to the concept of time, as certain interpretations suggest that time might not flow in the way we perceive it. For example, the **Many-Worlds Interpretation** of quantum mechanics

suggests that all possible events happen simultaneously in different "worlds," creating multiple timelines rather than a single, unified flow of time. This idea raises questions about the uniqueness of the present moment and the direction of time itself.

The nature of time thus occupies a unique place in metaphysical inquiry. Whether time is an external, universal progression or an internal construct shaped by consciousness, its implications touch on the very foundations of how we experience reality and construct meaning.

The Universe and Our Place Within It

The universe is the totality of all that exists, encompassing both space and time, and everything within them. From a metaphysical perspective, the universe is not merely a collection of stars, planets, and galaxies but a vast and interconnected whole that invites questions about origin, purpose, and our place within it.

1. **The Origins of the Universe:** One of the central questions in metaphysics concerns the origin of the universe. **Cosmology**, the study of the universe's origins and development, is a field that blends physics and metaphysics. The **Big Bang Theory** suggests that the universe began as a singularity, a point of infinite density and temperature that expanded to create space and time as we know them. However, the question of what preceded the Big Bang or why the universe came into being remains unanswered and may be beyond the scope of empirical science. Metaphysicians explore these questions by asking whether the universe has an underlying cause or purpose, and if it does, what that might be.

2. **Is the Universe Finite or Infinite?** Metaphysically, there are significant

implications in whether the universe is finite (with a boundary) or infinite (extending endlessly). If the universe is finite, it implies a structure or form with a limit, which raises questions about what, if anything, lies beyond it. If the universe is infinite, it challenges our understanding of space, as infinity is a concept that defies physical boundaries. The nature of the universe's size and shape leads to questions about its composition, origin, and possible end.

3. **Human Place in the Universe:** Metaphysical inquiry also examines humanity's place in this vast cosmos. Many philosophical traditions, from **ancient Greece** to **modern existentialism**, have speculated on whether the universe has a purpose and what role, if any, humans play within it. In some views, the universe is seen as purposeless, a vast and indifferent expanse within which humanity exists by chance. In other views, humans play a unique role

as conscious beings capable of reflection, morality, and spiritual insight.

4. **Cosmic Consciousness and Interconnectedness:** Some metaphysical perspectives propose that the universe is not only physical but also conscious, containing an underlying unity that connects all forms of existence. **Pantheism** suggests that the universe itself is divine, with everything connected through a universal consciousness or "God." Similarly, **panpsychism** argues that consciousness or experience may be a fundamental property of the universe, present even in seemingly inanimate matter. Such ideas suggest that humans are deeply interconnected with the cosmos, part of a larger, conscious whole.

5. **The Possibility of Multiple Universes:** Recent theories in physics and metaphysics suggest that our universe may be just one of many in a "multiverse." According to the **Many-Worlds Interpretation** of quantum mechanics and

other multiverse theories, every possible outcome of every event creates a new universe, leading to an infinite variety of universes with different physical laws, histories, and futures. This raises profound questions about our universe's uniqueness, the nature of reality, and our understanding of existence.

Mind and Matter

The relationship between mind and matter has fascinated philosophers for centuries, leading to profound questions about consciousness, existence, and reality itself. Mind and matter are two of the most central elements in metaphysics, where scholars explore whether these two realms are separate, interconnected, or even different manifestations of the same substance. This discussion often involves three main perspectives: **Dualism**, **Materialism**, and **Idealism**.

Dualism, Materialism, and Idealism

1. Dualism
Dualism posits that mind and matter are fundamentally distinct entities, each with

its own unique properties and laws. **René Descartes**, one of the most famous proponents of dualism, argued that the mind is a non-material, thinking substance, while the body is a material, extended substance. This perspective is often referred to as **Cartesian Dualism**, where the mind and body interact but remain separate entities.

- o **Substance Dualism**: This form of dualism holds that the mind and body are composed of different substances. The mind, or soul, is immaterial and non-physical, while the body and brain are made of physical matter. Substance dualism raises the "interaction problem," as it challenges us to understand how two fundamentally different substances can affect each other.

- o **Property Dualism**: Rather than asserting that mind and matter are separate substances, property dualism argues that the mind and brain are composed of the same substance (matter) but possess different properties. Here, the mind's properties, such as consciousness, thoughts, and emotions, are distinct and irreducible, meaning they cannot be explained entirely by physical processes.

○ **Criticisms of Dualism**: Many criticise dualism for failing to provide a clear mechanism for how non-physical consciousness could interact with the physical body. The lack of observable interactions between mind and matter has led some to view dualism as less scientifically plausible, pushing them toward other explanations like materialism.

2. **Materialism**

Materialism, sometimes referred to as **Physicalism**, argues that only matter exists and that everything, including consciousness and thoughts, can be explained through physical processes and interactions. Materialism asserts that the mind is not a separate substance but rather the result of complex processes in the brain.

- o **Reductionist Materialism**: In this view, mental states are reduced to or equated with physical states. For example, feelings of pain might be explained solely by neural activity in specific areas of the brain. This reductionist perspective seeks to explain consciousness in terms of neurobiology, suggesting that complex thoughts and emotions are simply the sum of brain processes.
- o **Emergent Materialism**: Emergent materialism posits that mental properties emerge from the brain's physical structure but cannot be entirely reduced to it. Here, consciousness is seen as an emergent property of complex

neural networks, one that appears only at certain levels of complexity and cannot be fully predicted by examining individual neurons alone.

- ○ **Criticisms of Materialism**: Critics argue that materialism struggles to explain subjective experience, often called the **"hard problem of consciousness"**. While materialism can describe brain functions, it has yet to fully explain why conscious experience exists at all—why we have a "first-person" perspective and feel emotions or sensations in the way we do.

3. **Idealism**

   Idealism takes the opposite approach to materialism, asserting that the mind is the only reality and that the material world is a construct of consciousness. **George Berkeley**, a notable idealist philosopher, argued that material objects do not exist independently of perception. For him, "to be is to be perceived," meaning that what we perceive as physical objects are ultimately mental constructs.

- o **Subjective Idealism**: In this form of idealism, reality exists only within the mind of the perceiver. There is no objective world; instead, all that exists is the individual's perception and mental constructs. This extreme view suggests that the world would cease to exist if there were no conscious beings to perceive it.

- o **Objective Idealism**: This view maintains that while individual perceptions are subjective, there exists a universal consciousness or mind that serves as the foundation for all existence. This universal mind gives coherence and continuity to reality, ensuring that it exists even when individual minds are not perceiving it.

- ○ **Criticisms of Idealism**: The main criticism of idealism is its challenge to common sense and empirical evidence. Since idealism posits that the physical world is a mental construct, it becomes difficult to account for the consistency and predictability of physical laws, which science relies on.

Each of these perspectives—dualism, materialism, and idealism—provides a unique framework for understanding the relationship between mind and matter. Each view raises questions and faces challenges, highlighting the complexity of this fundamental metaphysical issue.

The Relationship Between Mind and Body

The **mind-body problem** examines how the mind and body interact and influence one another. For centuries, this issue has challenged philosophers, especially dualists who argue for the separate nature of mind and body. Different perspectives offer varied explanations for this relationship:

1. **Interactionism**
   Interactionism is the view that the mind and body, although different in nature, causally interact. Descartes suggested that the **pineal gland** in the brain was the "seat of the soul," where mind and body influenced each other. Modern neuroscience, while not supporting Descartes' exact view, has observed that mental states such as thoughts and emotions can cause physiological responses in the body, such as increased heart rate during anxiety. This interaction suggests that mind and body can indeed affect each other, even if they are fundamentally different.

2. **Epiphenomenalism**
   Epiphenomenalism argues that while physical processes in the brain give rise to

consciousness, mental states are merely by-products of brain activity and do not have any causal power. In this view, thoughts and emotions arise from brain activity but cannot influence the brain or body. Critics argue that this fails to account for how mental states can seem to drive physical actions, like consciously deciding to move a hand.

3. **Parallelism**

   Parallelism suggests that mind and body are separate and do not interact, but that they move in parallel, synchronised by a pre-established harmony. This view, held by **Leibniz**, posits that God set up a harmony between mind and body at the creation of the universe, so they operate in sync without direct interaction. This theory sidesteps the need for physical interaction between mind and body but relies on a metaphysical synchronisation that many find implausible.

4. **Modern Neuroscientific Approaches**

   Advances in neuroscience have provided

insight into how brain processes correlate with mental states, supporting a materialist view of the mind-body connection. Brain imaging has shown that specific brain regions activate during particular thoughts or emotional states, leading to a deeper understanding of how consciousness may arise from physical processes. Despite this progress, a complete explanation of consciousness remains elusive, and the question of how subjective experience arises from physical processes continues to challenge scientists and philosophers alike.

How Thoughts Shape Reality

The idea that thoughts shape reality is central to metaphysics and finds expression in various traditions, from ancient philosophy to modern psychology. This concept suggests that our perception, beliefs, and expectations can influence our experience of the world in profound ways.

1. **Perception and Reality**
   How we perceive the world often shapes our experience of it. For example, two individuals in the same situation may have different interpretations based on their beliefs and attitudes. This subjective filtering of experiences indicates that reality is partly constructed by the mind. Psychologists refer to this as **cognitive bias**, where personal expectations and preconceptions shape how we interpret events.

2. **The Power of Belief and Self-Fulfilling Prophecies**
   A **self-fulfilling prophecy** occurs when our beliefs or expectations influence our behaviour in a way that makes those beliefs come true. For instance, someone who believes they are confident may behave in a way that leads others to perceive them as such, reinforcing their

confidence. This shows how thoughts and beliefs can bring about tangible changes in reality.

3. **Quantum Physics and the Observer Effect**
Quantum mechanics has introduced the idea that the act of observation itself can influence physical reality. The **observer effect** in quantum physics suggests that particles can exist in multiple states until observed, at which point they "collapse" into a single state. This phenomenon has led some to speculate that consciousness plays a role in shaping reality, though this interpretation remains highly controversial and debated.

4. **Manifestation and the Law of Attraction**
The **Law of Attraction** posits that positive or negative thoughts bring positive or negative experiences into one's life. According to this belief, focusing on desired outcomes can influence one's actions and attitudes, which in turn attract

those outcomes. While scientifically unproven, the law of attraction is popular in self-help and metaphysical circles, emphasising the power of thoughts to shape life circumstances.

5. **Consciousness as a Creative Force**
   Some metaphysical traditions view consciousness itself as a creative force, suggesting that reality is shaped by the collective thoughts, emotions, and beliefs of conscious beings. This perspective aligns with ideas in **pantheism** and **panpsychism**, where consciousness is seen as a fundamental component of reality rather than a byproduct of the brain. In this view, our thoughts do not merely reflect reality; they actively participate in its creation.

The Concept of the Soul

The concept of the soul is central to many philosophical, religious, and metaphysical discussions, representing the essence or true self that exists beyond the physical body. Throughout history, the soul has been depicted as the core of identity, the repository of consciousness, and the source of moral and spiritual insight. Though definitions vary widely across cultures and belief systems, the soul is generally seen as an immortal or eternal aspect of human existence, one that transcends physical limitations and persists beyond death.

What is the Soul? Different Philosophical Views

1. The Classical Perspective

In ancient philosophy, especially among Greek thinkers, the soul was often considered the essence of life and the animating principle behind all living beings. **Plato** famously viewed the soul as the immortal and immutable essence of a person, distinct from the body and capable of existing apart from it. For Plato, the soul pre-exists the body and returns to a divine realm upon death, where it contemplates pure forms or ideals. This belief in the immortality of the soul and its separation from the body profoundly influenced later Western thought. **Aristotle**, however, took a different approach. For him, the soul (or *psyche*) was the "form" or "essence" of the body, but it was not independent or separable from it. According to Aristotle, the soul gives life to the body but does not continue once the body dies. Unlike Plato, he believed that the soul and body are interdependent, much like the relationship between a shape and the substance that takes it.

2. **Dualism and the Cartesian View**
   René Descartes argued that the soul, or mind, is entirely separate from the physical body, embodying the principle of **dualism**. For Descartes, the soul is a thinking substance—*res cogitans*—which is distinct from the body, or *res extensa*, the extended or physical substance. This view sees the soul as the true self, an immaterial entity that interacts with the body but exists independently from it. Descartes' dualism became a foundation for much Western thought on the nature of consciousness and personal identity.

3. **Materialism and Reductionism**
   Materialist and reductionist philosophers argue that what we call the soul is merely the product of complex physical processes in the brain and nervous system. For

materialists, there is no separate, non-physical soul; consciousness, identity, and personality all arise from brain functions. Once these brain functions cease, consciousness ceases as well, leaving no enduring essence or soul.

4. **Idealism and Spiritual Monism**
   Idealist philosophers, such as **George Berkeley** and many in Eastern traditions, argue that the soul or mind is the primary reality, and the physical world may be an extension or manifestation of consciousness. In this view, the soul is not an isolated entity but an integral part of a larger spiritual reality or divine consciousness. In **Advaita Vedanta** (a school of Hindu philosophy), for instance, the individual soul, or *atman*, is ultimately identical with *Brahman*, the universal spirit, suggesting that all individual souls are connected within a greater, singular consciousness.

Theories of Life, Death, and the Afterlife

1. **Eternal Life and Immortality**
   Many religious and spiritual traditions
   believe that the soul is inherently
   immortal. Christianity, Islam, and Judaism
   traditionally hold that the soul persists
   after death, facing judgement or
   transformation according to one's deeds in
   life. In Christianity, for example, the soul
   either ascends to heaven, descends to hell,
   or awaits resurrection, depending on
   divine judgement. This belief in eternal
   life reflects the idea that the soul is sacred
   and that its ultimate purpose is to attain a
   form of divine communion or moral
   perfection.

2. **Annihilation and Non-Existence**
   Some perspectives suggest that the soul
   may not survive death or that
   consciousness simply ceases when the
   body dies. Philosophers who support
   annihilationism argue that there is no
   evidence for an afterlife and that

consciousness, being a product of the brain, ends with physical death. Some Eastern philosophies, like **Theravada Buddhism,** suggest that the self is an illusion (anatta), and rather than a permanent soul, what persists after death are the effects of karma, while personal identity dissolves.

3. **Resurrection and Rebirth**
Resurrection involves the reanimation or transformation of the soul into a new form or body after death. In Christianity, resurrection is a central theme, promising believers that their souls and bodies will be transformed in the afterlife. The idea is that the soul retains its identity but is freed from earthly limitations, experiencing life in a perfected or transcendent form.

4. **Soul Journey and Ascension**
Some traditions view death as the beginning of a journey where the soul undergoes transformation and enlightenment. In **Egyptian mythology,** for example, the soul embarks on a

perilous journey through the underworld, facing trials before reaching a peaceful afterlife. This journey reflects the belief that death is a process of growth, purification, or transition, rather than an end. In **Theosophy** and certain New Age beliefs, the soul's ascension is viewed as part of a cosmic evolution toward higher states of consciousness, with each lifetime serving as a lesson for spiritual advancement.

Reincarnation, Karma, and Eternal Consciousness

1. **Reincarnation**
   Reincarnation is the belief that the soul is reborn in new bodies over successive lifetimes, continuing a cycle of birth, death, and rebirth. This belief is foundational in **Hinduism, Buddhism, Jainism**, and **Sikhism**, where the soul undergoes countless lifetimes as it

progresses toward enlightenment or
liberation (*moksha* in Hinduism and
nirvana in Buddhism).

- ○ **Hindu View**: In Hindu philosophy,
    the soul (or *atman*) moves through
    different bodies based on the effects
    of karma. The ultimate goal is to
    escape the cycle of rebirth
    (samsara) and unite with *Brahman*,
    the universal consciousness.
- ○ **Buddhist View**: Although
    Buddhism doesn't assert a
    permanent soul, it believes in the
    continuation of consciousness
    across lifetimes through the cycle of
    samsara. Rebirth continues until
    one attains enlightenment and
    reaches *nirvana*, freeing oneself
    from the suffering of existence.
- ○ **Western Interpretations**: In
    Western contexts, reincarnation is
    less common in mainstream
    religions but has gained interest in
    New Age and esoteric movements,

where it's seen as a way for souls to learn and evolve spiritually.

2. **Karma**

Karma is the principle that one's actions influence future experiences, shaping future lives and circumstances. Originating in Eastern thought, karma operates as a cosmic law of cause and effect, guiding the soul's journey through various incarnations.

- o **Moral Responsibility**: In Hindu and Buddhist traditions, karma is both an ethical and cosmic principle. Good actions lead to positive outcomes, while negative actions result in suffering. Karma doesn't only affect the individual in this life but carries over to influence future incarnations.

- o **Learning and Growth**: For many believers in reincarnation, karma is not simply a reward-punishment system but a way for the soul to learn. Each lifetime offers new opportunities to gain wisdom and overcome weaknesses, helping the soul advance toward enlightenment or liberation.

3. **Eternal Consciousness**

Some metaphysical perspectives propose that consciousness is not bound to individual bodies but is instead a part of a vast, eternal reality. **Panpsychism** and **pantheism** suggest that all of existence is imbued with a form of consciousness or spirit.

- ○ **Universal Mind**: This concept argues that all individual consciousnesses are expressions of a universal or cosmic mind. Some interpret this as a form of **collective consciousness**, where individual souls are interconnected parts of a greater whole.
- ○ **Continuity Beyond Individuality**: The idea of eternal consciousness emphasises that personal identity may dissolve after death, yet

consciousness continues in some form, beyond individual experiences. In some **New Age** and **esoteric** beliefs, individuals can attain a state of universal consciousness, experiencing unity with all of existence and moving beyond the limitations of ego and personal identity.

4. **Modern Scientific Speculation**

   While science has yet to confirm the existence of the soul, some researchers explore consciousness as a fundamental property of the universe. **Quantum consciousness** theories, for instance, suggest that consciousness might have a basis in quantum mechanics, which could imply that consciousness is not entirely tied to the physical body. **Near-death experiences (NDEs)** and **out-of-body experiences (OBEs)** have also led to ongoing research and debate about consciousness and the possibility of

survival after death, although scientific
evidence remains inconclusive.

Causality and Free Will

The interplay between causality and free will has captivated thinkers for centuries, particularly within metaphysics, where understanding the true nature of human choice and causation remains a core challenge. At its essence, causality explores the relationship between causes and effects—underlying principles that govern change, movement, and progress. Free will, on the other hand, delves into the possibility that humans may have autonomy over their actions, shaping their lives independently of deterministic forces. This dynamic poses complex questions about whether our lives are predetermined or if we possess genuine agency.

The Principle of Cause and Effect

Causality, or the principle of cause and effect, suggests that every event or occurrence is the result of a prior cause. This principle is foundational in both science and philosophy and forms the basis of many metaphysical inquiries about the nature of reality, change, and time.

1. **Understanding Causation in Metaphysics**

   Metaphysicians have long debated the exact nature of causation. In simple terms, causation implies that for every effect, there is a corresponding cause that brought it into being. This concept is apparent in everyday life, where actions or events seem to follow from other actions or events predictably. For instance, lighting a match causes it to ignite, a stone thrown into water creates ripples, and so forth. This relationship between causes

and effects forms a chain, suggesting that
all actions are interlinked.

2. **Types of Causes**

   Aristotle identified four kinds of causes:

   ○ **Material Cause**: What something
   is made of (e.g., a statue made of
   marble).

   ○ **Formal Cause**: The structure or
   form of a thing (e.g., the shape of
   the statue).

   ○ **Efficient Cause**: The agent or force
   that brings something into being
   (e.g., the sculptor who carves the
   statue).

   ○ **Final Cause**: The purpose or goal
   of a thing (e.g., the statue's intended
   beauty or function as art).

3. These categories provide a broader
   understanding of causation, encompassing
   not just physical events but the purposes
   and intentions behind them, thus merging
   objective and subjective aspects of
   causation.

4. **Causality and the Chain of Events**
   Causality suggests that the universe operates as an interconnected chain of events where each effect becomes the cause of another effect. This notion leads to the principle of **causal determinism**, which posits that every present state or event is the inevitable outcome of preceding states and causes. In this view, the universe is a closed system where everything unfolds according to predictable laws. This deterministic framework leaves little room for randomness or genuine freedom, prompting questions about whether free will can coexist with an unbroken causal chain.

5. **Quantum Mechanics and Causation**
   The advent of **quantum mechanics** challenged traditional views of causality, introducing a level of unpredictability at the subatomic level. Unlike classical physics, which operates on fixed laws of cause and effect, quantum mechanics

suggests that particles may behave randomly or probabilistically. This has led some to speculate that randomness or indeterminacy at the quantum level could allow for a kind of "freedom" from strict causation, which may provide a foundation for free will. However, the relationship between quantum indeterminacy and human freedom remains controversial, as the randomness in quantum mechanics doesn't directly translate to conscious decision-making.

Determinism vs. Free Will

The debate between determinism and free will lies at the heart of metaphysical inquiry, particularly regarding human autonomy. Determinism suggests that all events, including human actions, are determined by prior causes and that given the initial conditions of the universe, everything that happens is predictable. Free will, conversely, proposes that humans have the capacity to make choices that are not fully determined by external circumstances.

1. **Determinism**

   Determinism asserts that every action and event in the universe, including thoughts and decisions, is the inevitable result of previous causes. **Hard determinism** takes this to mean that free will is entirely an illusion; we may feel as though we're making independent choices, but in reality, every decision is governed by factors beyond our control, such as genetics, environment, and past experiences.

   ○ **Scientific Determinism**: Rooted in physics and biology, scientific determinism suggests that physical laws govern every process, including human actions. For example, a person's genetic makeup and environmental conditioning shape their behaviour in predictable ways, with little room for autonomous choice.

- **Psychological Determinism**: Psychological determinism, proposed by thinkers like **Sigmund Freud**, argues that unconscious desires and early experiences largely dictate human behaviour. From this viewpoint, people may have little true freedom, as their choices are directed by internal psychological drives formed by previous experiences.

2. **Free Will**

Free will is the belief that individuals possess a degree of autonomy that allows them to make choices independently of external causes. Supporters of free will argue that even if certain aspects of behaviour are influenced by genetic or environmental factors, humans can still exercise a level of personal agency.

- **Libertarian Free Will**: This position asserts that individuals can initiate actions independent of deterministic causes. In this view, free will is genuine and not an illusion, and humans are morally responsible for their choices.

- ○ **Compatibilism**: Compatibilists believe that determinism and free will can coexist. They argue that free will does not require absolute independence from causation but rather the ability to act according to one's desires and reasons, even if these are shaped by prior causes. Compatibilism suggests that people can be both determined and free, as long as they act in alignment with their internal motivations.

3. **Influential Thinkers on Determinism and Free Will**
   - ○ **Immanuel Kant**: Kant attempted to reconcile determinism and free will, suggesting that while we experience the physical world as causally determined, human actions are

governed by moral laws beyond the empirical world. He saw free will as essential for moral responsibility, arguing that we must assume freedom when considering moral choices.

- **Jean-Paul Sartre**: Sartre argued from an **existentialist** perspective that humans have radical freedom and are fully responsible for their choices. In his view, individuals are "condemned to be free," meaning they must create their essence through actions in an otherwise meaningless world.

How Metaphysics Views Choice and Destiny

Metaphysics offers various perspectives on the relationship between choice, destiny, and free will. These views range from fatalism to the possibility of self-determined freedom, providing nuanced approaches to understanding human agency and its limitations.

1. **Fatalism**
   Fatalism is the belief that events are fixed in advance and that human choices cannot alter their course. In a fatalistic worldview, everything unfolds according to a preordained destiny, suggesting that choice is merely an illusion. This perspective is common in certain religious traditions, where divine will or cosmic law predetermines all events.

2. **Predestination and Divine Will**
   In some religious doctrines, such as certain branches of **Christianity** and **Islam**, the concept of predestination implies that God has foreordained every event, including human choices. This belief posits that destiny and free will coexist under divine sovereignty; humans appear to make independent choices, but these choices are ultimately within God's predetermined plan.

3. **Existentialism and Radical Freedom**

   Existentialist philosophers argue that individuals are entirely free and must create their path in life. For **Sartre** and others, freedom is an absolute, and with it comes full responsibility for one's actions. This form of radical freedom emphasises the role of personal choice in determining one's destiny, suggesting that each individual constructs meaning and purpose through their decisions.

4. **The Role of Karma in Eastern Philosophy**

   In Eastern philosophies like **Hinduism** and **Buddhism**, karma shapes destiny, but individuals still retain freedom to act. Karma suggests that one's past actions influence current and future circumstances, yet it allows room for free will, enabling individuals to choose actions that lead to liberation or further attachment. This balance acknowledges causality without denying personal agency, emphasising growth, responsibility, and self-determination.

5. **The Multiverse Theory and Alternate Choices**

   Metaphysical interpretations of modern physics, like the **multiverse theory**, suggest that every possible outcome of a decision exists in some alternate universe. This implies that choice could create multiple, coexisting realities where each decision leads to a different outcome. This approach allows for both causality and free will by proposing that all choices are "realised" in different branches of the multiverse.

6. **Metaphysical Implications of Choice and Destiny**

   Metaphysics often addresses the paradox of destiny and choice by proposing that the nature of reality may be more complex than a simple deterministic framework. For example, **process philosophy** suggests that reality is in a constant state of becoming and that human consciousness and choices contribute to the unfolding of reality. This view allows for the idea that humans are both bound by causation and capable of affecting the future, emphasising a dynamic, interconnected reality where individual choices have meaningful consequences.

Energy and the Universe

The concept of energy is fundamental to both physics and metaphysics, serving as a bridge between the tangible and intangible aspects of existence. While in physics, energy is defined in terms of measurable properties like kinetic energy, potential energy, and thermodynamics, metaphysics delves deeper, exploring energy as a more abstract and essential element that permeates the universe. Understanding energy beyond its physical definitions invites inquiry into the vibrational frequencies that underpin all existence and the interconnectedness of all things.

Understanding Energy Beyond the Physical

1. **Defining Energy**

   In physics, energy is often quantified and categorised into forms, such as kinetic energy (the energy of motion), potential energy (stored energy based on position), thermal energy, and electromagnetic energy. However, from a metaphysical perspective, energy transcends these classifications, representing a universal force that fuels not only physical processes but also the underlying essence of existence.

2. **Energy as a Universal Principle**

   In metaphysical thought, energy is viewed as a universal principle that connects all aspects of reality. The idea that everything in the universe is made of energy can be traced back to ancient philosophies, including those found in Eastern traditions like **Taoism** and **Buddhism**, where concepts like **Qi** (or Chi) and **Prana** refer to the vital life force that animates all living beings and connects them to the universe. This perspective implies that energy is not just a physical phenomenon but also a spiritual essence that binds the cosmos together.

3. **The Nature of Energy**
Energy can be perceived in various forms beyond the physical, such as emotional, spiritual, and psychological energy. These forms influence human experiences, relationships, and overall well-being. For instance, positive emotions like love and joy are thought to generate high-frequency energies, while negative emotions such as fear and anger are associated with lower frequencies. This understanding aligns with the idea that energy shapes our interactions with the world and with each other.

4. **Quantum Physics and Energy**
Quantum physics has significantly impacted our understanding of energy at a fundamental level. The **wave-particle duality** of light and matter suggests that

all particles can behave as both waves and particles, highlighting the energy's inherent dual nature. Moreover, the famous equation $E=mc^2$, proposed by **Albert Einstein**, demonstrates that energy and mass are interchangeable, reinforcing the idea that energy is a fundamental component of reality. This has profound implications for metaphysical inquiries, suggesting that everything in the universe is interconnected through a shared energy fabric.

Exploring Vibrational Frequencies

1. **The Nature of Vibrational Frequencies**
   At the core of metaphysical thought about energy is the idea of vibrational frequencies. All matter, from the smallest particle to the largest celestial body, vibrates at specific frequencies. These frequencies determine the properties and behaviour of matter, influencing how it

interacts with other forms of energy. Vibrational frequencies are not just limited to the physical realm; they extend into emotional and spiritual dimensions, resonating with the belief that everything in the universe is in a state of constant motion and vibration.

2. **Resonance and Harmony**

   The concept of resonance describes how objects or systems can amplify or diminish each other's vibrations when they interact. In metaphysical terms, this notion suggests that individuals and environments can resonate with specific frequencies, affecting their well-being and experiences. For example, practices like **sound healing** use instruments such as singing bowls or tuning forks to create sound vibrations that can restore balance and harmony within the body, mind, and spirit. This resonance theory connects to the idea that our thoughts, emotions, and intentions can influence our personal vibrational frequencies.

3. **Meditation and Frequency Alteration**
   Practices such as meditation, mindfulness, and various forms of energy work aim to alter one's vibrational frequency to achieve a higher state of consciousness. Through focused intention and practice, individuals can raise their vibrational frequencies, leading to experiences of greater clarity, peace, and connection with the universe. Many spiritual traditions assert that raising one's frequency aligns individuals with higher states of being, unlocking deeper insights and spiritual awareness.

4. **The Law of Attraction**
   The **Law of Attraction** posits that like attracts like, suggesting that individuals attract experiences, people, and opportunities that resonate with their vibrational frequencies. According to this principle, maintaining a positive and high-frequency mindset can draw similar energies into one's life, while negative thoughts and feelings may attract discord

and challenges. This perspective emphasises the power of individual consciousness in shaping personal realities, fostering an understanding of how vibrational frequencies play a crucial role in manifesting experiences.

The Interconnectedness of All Things

1. The Web of Life

The interconnectedness of all things is a fundamental tenet of metaphysical thought. This principle posits that every being, object, and phenomenon in the universe is linked by an intricate web of energy. The idea of **interdependence** suggests that changes in one part of this web can have ripple effects throughout the entire system. This interconnectedness resonates with ecological principles, where the health of one aspect of an ecosystem affects the whole, but it also extends to human relationships, social structures, and spiritual connections.

2. **The Holographic Universe**
   Some metaphysical perspectives propose that the universe operates like a hologram, where each part contains the information of the whole. This theory suggests that every individual is a reflection of the larger universe, implying that understanding oneself can lead to insights about the cosmos. This notion resonates with the idea that individual actions and thoughts can influence the collective, reinforcing the importance of consciousness in shaping reality.

3. **Collective Consciousness**
   The idea of a **collective consciousness** suggests that all human beings are interconnected through a shared consciousness, impacting thoughts, beliefs, and emotions on a global scale. This concept is echoed in various spiritual teachings and psychological theories, emphasising how societal values and

cultural norms can influence individual behaviours. Understanding this interconnectedness can lead to greater empathy and awareness, encouraging individuals to recognize their role within the larger fabric of existence.

4. **Energy Exchange and Relationships**
   The interconnectedness of all things is especially evident in human relationships, where energy is exchanged constantly. The interactions between individuals create an energetic dynamic that influences emotional states, mental clarity, and overall well-being. Recognizing this exchange can empower individuals to cultivate healthier relationships, as they become aware of how their energy impacts others and vice versa. Practices such as active listening, compassion, and presence can enhance these exchanges, promoting positive vibrational frequencies in interpersonal connections.

5. **Spiritual Growth and Interconnectedness**

Embracing the idea of interconnectedness can lead to profound spiritual growth. As individuals recognize their connection to all beings, they may cultivate a sense of unity and compassion, transcending boundaries of individuality. This shift in perception fosters a deeper understanding of shared experiences and mutual support, ultimately leading to a more harmonious existence.

The Metaphysics of Knowledge

The metaphysics of knowledge, often encapsulated within the branch of philosophy known as **epistemology**, seeks to understand the nature, scope, and limits of human knowledge. It addresses fundamental questions about how we know what we know, the sources and validity of our beliefs, and the relationship between knowledge and reality.

Epistemology: How Do We Know What We Know?

1. **Defining Epistemology**
   Epistemology is the philosophical study of

knowledge, its nature, sources, limitations, and validity. It probes questions such as: What constitutes knowledge? How is knowledge acquired? What is the distinction between belief and knowledge? Traditionally, knowledge has been defined as **justified true belief**, suggesting that for someone to claim to know something, it must be true, they must believe it, and there must be sufficient justification for that belief.

2. **The Sources of Knowledge**
   Epistemologists have identified several sources of knowledge, including:
   - **Perception**: Our senses provide us with empirical data about the world. However, perception can be deceptive, leading to philosophical scepticism about whether we can truly know anything based on sensory experience alone.
   - **Reason**: Rationalism posits that knowledge can be acquired through reason and logical deduction.

Thinkers like **René Descartes** emphasised the role of reason as a primary source of knowledge, proposing that certain truths are innate and can be discovered through introspection.

○ **Testimony**: Much of what we know comes from the accounts of others. This raises questions about the reliability of sources and the role of authority in shaping our beliefs.

○ **Intuition**: Some epistemologists argue that knowledge can also arise from intuitive insights, which provide immediate understanding without the need for conscious reasoning or empirical evidence.

3. **The Problem of Scepticism**

One of the central challenges in epistemology is scepticism, which questions whether knowledge is possible at all. Sceptics argue that our senses can deceive us, and thus, we cannot have absolute certainty about our beliefs. This has led to various responses in the field, such as **fallibilism**, which acknowledges that while certainty may be unattainable, we can still hold justified beliefs that can guide our actions and understanding of the world.

4. **Constructivism and Relativism**
   Epistemological constructivism posits that knowledge is not merely discovered but constructed by individuals or societies based on their experiences and contexts. This perspective suggests that knowledge is subjective and may vary across different cultures and historical periods. Relativism further challenges the notion of objective truth, positing that what is considered knowledge is relative to specific frameworks or belief systems, thereby complicating the search for universal truths.

Intuition, Reason, and Perception

1. **The Role of Intuition**

   Intuition is often described as a form of immediate knowledge or understanding that arises without conscious reasoning. This form of knowledge can be particularly powerful, as it often guides individuals in decision-making processes or creative endeavours. **Philosophers like Immanuel Kant** argued that intuition is a critical component of knowledge, especially in understanding concepts that are beyond empirical observation, such as morality or metaphysical truths.

   - **Intuitive Knowledge**: Intuition can manifest in various ways, including gut feelings, insights during meditation, or sudden realisations. It is often contrasted with analytical reasoning, which involves systematic and logical processing of information.

2. **Reason as a Tool for Knowledge**

   Reason plays a pivotal role in shaping our understanding of the world. Rationalism posits that human beings possess innate faculties of reason that allow them to arrive at truths independent of sensory experience. Philosophers such as **Gottfried Wilhelm Leibniz** and **Baruch Spinoza** emphasised the significance of reason in the pursuit of knowledge, advocating for a systematic approach to philosophical inquiry.

- **The Limitations of Reason**: While reason is a powerful tool, it is not infallible. Logical fallacies, biases, and assumptions can cloud rational thinking, leading to flawed conclusions. Understanding these limitations is crucial for developing a more nuanced epistemology.

3. **Perception and Knowledge**

Perception serves as one of the primary avenues through which we gather knowledge about the external world. However, it is fraught with challenges. Our sensory experiences can be misleading, leading to philosophical debates about the nature of reality and our ability to perceive it accurately.

- ○ **Phenomenology**: The study of how things appear to consciousness, phenomenology examines the subjective experience of perception and its implications for knowledge. Philosophers like **Edmund Husserl** and **Maurice Merleau-Ponty** explored how our embodied experiences shape our understanding of the world.
- ○ **Theories of Perception**: Various theories explain how perception contributes to knowledge, including direct realism (which asserts that we perceive the world directly as it is) and representationalism (which suggests that our perceptions are mental representations of external reality).

Exploring Mysticism and Insight

1. **Mysticism and the Limits of Knowledge**
   Mysticism often concerns itself with direct, experiential knowledge of the divine or the ultimate nature of reality. Mystical experiences transcend ordinary perception and reason, offering insights that defy conventional understanding. **Mystics** throughout history, such as **Rumi**, **Buddha**, and **St. John of the Cross**, have described profound states of unity with the cosmos, emphasising the inadequacy of language and rational thought to fully capture these experiences.

2. **Intuitive Insight**
   Mysticism and intuition share common ground in their emphasis on non-rational ways of knowing. Intuitive insight often leads to moments of clarity that can feel revelatory, suggesting that there are dimensions of knowledge that lie beyond empirical evidence and logical deduction. Such insights can foster a deeper understanding of life's mysteries, inviting

individuals to explore their consciousness more fully.

3. **The Intersection of Mysticism and Epistemology**
The relationship between mysticism and epistemology raises intriguing questions about the nature of knowledge itself. Can mystical experiences provide legitimate knowledge? Are they to be considered subjective insights or valid forms of understanding? These inquiries encourage a reevaluation of the boundaries of epistemology, suggesting that alternative modes of knowing should be acknowledged and integrated into the broader discourse.

4. **Meditative Practices and Insight**
Many spiritual traditions emphasise the importance of meditation and contemplative practices as means of gaining insight. These practices often facilitate a shift from analytical thinking to a state of heightened awareness, allowing individuals to tap into deeper

layers of consciousness. Such insights can lead to transformative experiences that fundamentally alter one's perception of self and reality.

God and the Divine

The exploration of God and the divine is a profound aspect of metaphysical inquiry, delving into the fundamental nature of existence, reality, and the ultimate source of all that is. Different philosophical traditions and religious beliefs offer varied perspectives on the divine, leading to rich discussions surrounding the existence of God, the attributes ascribed to divinity, and the implications of these beliefs for understanding the universe. Within this framework, metaphysical thought grapples with essential questions: What is the nature of God? How do we know of the divine? And what role does divinity play in the overarching tapestry of existence?

The Existence of God: Metaphysical Perspectives

1. **Arguments for the Existence of God**
   Various metaphysical arguments have been proposed to affirm the existence of God, each stemming from different philosophical traditions. Among the most prominent are:
   - **The Cosmological Argument**: This argument posits that everything that exists has a cause, and tracing these causes back leads to an uncaused cause, which is identified as God. Thinkers such as **Thomas Aquinas** articulated this in his "Five Ways," emphasising that the existence of the universe necessitates a first cause beyond itself.
   - **The Teleological Argument**: Also known as the argument from design, it suggests that the order and complexity observed in the

universe imply an intelligent designer. Advocates, including **William Paley**, argued that just as a watch implies a watchmaker, the intricate workings of nature point to a divine creator.

○ **The Ontological Argument**: Proposed by philosophers like **Anselm of Canterbury** and later expanded by **Descartes**, this argument asserts that God's existence is inherent to the very concept of God. If we can conceive of a being greater than which nothing can be conceived, it must exist, as existence is a necessary attribute of a perfect being.

○ **The Moral Argument**: This posits that the existence of moral values and duties suggests a moral lawgiver. Philosophers like **Immanuel Kant** argued that the universal sense of right and wrong indicates an underlying divine presence that transcends human subjectivity.

2. **Atheistic Perspectives**

   In contrast, atheistic philosophies challenge the necessity of a divine being. Prominent thinkers, such as **Friedrich Nietzsche** and **David Hume**, have argued against traditional proofs of God's existence. Hume questioned the reliability of human reason in grasping the divine, while Nietzsche famously declared the "death of God," suggesting that humanity must forge its own values in the absence of a divine moral authority.

3. **Agnosticism and the Limits of Knowledge**

   Agnosticism posits that the existence of God is unknowable or currently unknown. Agnostic philosophers argue that human beings lack the capacity to comprehend or prove the existence of the divine. This perspective emphasises the limits of human reason and experience, suggesting that metaphysical inquiry must acknowledge uncertainty regarding the divine.

4. **Experiential Evidence**

   Many individuals report personal experiences of the divine, which serve as a form of evidence for the existence of God. Mystical experiences, profound moments of connection with the universe, and feelings of transcendence are often cited as indicators of a higher power. Such subjective experiences challenge the strictly rational approach to understanding God, indicating that personal encounters with the divine may constitute a legitimate form of knowledge.

The Role of Divinity in Metaphysical Thought

1. **Divinity as the Ground of Being**

   In many metaphysical frameworks, God is conceptualised as the fundamental ground of being. **Heidegger**, for example, emphasises the concept of Being itself as the source from which all existence emanates. In this view, God or the divine is not merely an entity among entities but rather the very essence of existence, grounding all reality and imbuing it with meaning.

2. **Divine Attributes**

   Philosophers often attribute various characteristics to God, shaping their metaphysical understanding of the divine. Common attributes include:

   o **Omnipotence**: The idea that God is all-powerful, capable of doing anything logically possible.

   o **Omniscience**: The belief that God possesses all knowledge, encompassing the past, present, and future.

   o **Omnibenevolence**: The attribute of God as all-good, implying a divine nature that inherently desires the well-being of creation.

   o **Immutability**: The belief that God is unchanging and eternal, standing apart from the flux of the physical world.

3. **God and the Problem of Evil**
The existence of evil poses significant challenges to traditional theistic views. Theodicies attempt to reconcile God's goodness and omnipotence with the presence of suffering and evil in the world. Philosophers like **Augustine** and **Leibniz** proposed that evil arises not from God but from human free will or the necessity of a greater good that can emerge from suffering.

4. **The Divine and Human Experience**
The role of divinity extends to human experience and consciousness. Many metaphysical traditions propose that human beings possess a spark of the divine within them, leading to an understanding of God as immanent and transcendent. This duality suggests that while God is beyond human comprehension, divine qualities manifest

in the world through love, creativity, and consciousness.

Exploring Pantheism, Monotheism, and Polytheism

1. **Monotheism**

   Monotheism is the belief in a singular, all-powerful God. This perspective is central to major world religions such as **Judaism, Christianity,** and **Islam**. Monotheistic thought emphasises the uniqueness of the divine and often posits a personal God who interacts with creation. Key attributes of monotheistic belief include:

   - **Covenantal Relationship**: Many monotheistic traditions describe a covenant between God and humanity, emphasising ethical conduct and devotion.
   - **Revelation**: Monotheistic religions often assert that God reveals truths

about Himself and the nature of existence through sacred texts and prophets.

2. **Polytheism**

In contrast, polytheism acknowledges multiple gods, each with distinct roles, attributes, and domains. Ancient religions, such as those of **Greece, Rome,** and **Egypt**, illustrate how societies have understood the divine through diverse deities that embody various aspects of existence—love, war, agriculture, etc. This multiplicity allows for a rich tapestry of narratives and cultural practices centred on divine worship.

- **Mythological Frameworks**: Polytheistic belief systems often incorporate myths that explain natural phenomena, human existence, and moral dilemmas. These narratives provide a means for understanding the world and humanity's place within it.

3. **Pantheism**

   Pantheism posits that God and the universe are one and the same, rejecting the notion of a distinct creator separate from creation. This perspective suggests that the divine is immanent in all things, emphasising the sacredness of the natural world. Thinkers like **Baruch Spinoza** articulated pantheistic views, suggesting that understanding God requires a deep engagement with nature and reality.

   o **Interconnectedness of All Things**: Pantheism emphasises the interconnectedness of all existence, suggesting that recognizing this unity leads to a greater appreciation for life and a sense of reverence for the universe. In this view, every element of existence reflects the divine, fostering an understanding of spirituality that is rooted in the material world.

4. **Panentheism**

   Related to pantheism, panentheism posits

that while God encompasses the universe, He also transcends it. This perspective acknowledges a personal, relational aspect of God while asserting that the divine is not limited to the physical world. Panentheism allows for a dynamic relationship between the divine and creation, suggesting that while God is present in all things, He also exists beyond them.

Ethics and Morality in Metaphysics

Ethics and morality form a crucial component of metaphysical inquiry, addressing the fundamental questions surrounding right and wrong, good and evil, and the principles that guide human behaviour. The interplay between metaphysics and ethics offers profound insights into the nature of existence, the fabric of human interactions, and the overarching principles that govern moral decision-making. Through examining the origins of ethical thought, the metaphysical influences on moral principles, and the nature of good and evil, we can deepen our understanding of morality's role in human life and the universe.

The Origins of Right and Wrong

1. **Natural Law Theory**
   The concept of natural law suggests that moral principles are derived from the natural order of the universe. Philosophers like **Aristotle** and **Thomas Aquinas** argued that human beings possess an inherent sense of right and wrong, rooted in human nature and the rational understanding of the world. According to this view, moral laws reflect the intrinsic nature of reality, with ethical behaviour aligning with the natural purposes of human existence.
   - **Universal Morality**: Natural law theorists maintain that certain moral truths are universal, transcending cultural and societal boundaries. This universality implies that

ethical principles can be discovered through reason and reflection on the natural world.

2. **Cultural Relativism**

In contrast to natural law theory, cultural relativism posits that moral values are shaped by cultural contexts. This perspective emphasises that what is considered right or wrong varies across societies and historical periods. Proponents of cultural relativism argue that ethical norms are not fixed but are instead influenced by social, cultural, and historical factors.

   ○ **Ethical Pluralism**: This view recognizes the diversity of moral beliefs and practices across different cultures, advocating for tolerance and understanding of varying ethical systems. It also challenges the notion of an absolute moral truth, suggesting that morality is context-dependent.

3. **Moral Realism vs. Anti-Realism**
   The debate between moral realism and
   anti-realism centres around the existence
   of moral facts. Moral realists argue that
   ethical statements can be objectively true
   or false, similar to factual claims about the
   world. They assert that moral truths exist
   independently of individual beliefs or
   cultural contexts.

   - **Moral Anti-Realism**: Conversely,
     moral anti-realists maintain that
     moral values are subjective,
     dependent on individual
     perspectives or societal
     conventions. This view questions
     the objectivity of ethical claims,
     suggesting that morality is
     constructed rather than discovered.

4. **Divine Command Theory**
   Many religious traditions assert that moral
   principles originate from divine
   commands. According to this theory, what
   is considered right or wrong is determined
   by God's will, with ethical guidelines

outlined in sacred texts and religious teachings. This perspective raises questions about the relationship between divinity and morality, exploring how the divine influences ethical behaviour.

- ○ **The Euthyphro Dilemma**: This philosophical challenge, originating from **Plato**, questions whether moral values are good because God commands them or if God commands them because they are inherently good. This dilemma has led to extensive debates regarding the relationship between divinity and ethical principles.

How Metaphysics Influences Moral Principles

1. **Metaphysical Foundations of Morality** Metaphysics provides the foundational concepts that underpin ethical thought. For instance, notions of existence, being, and the nature of reality influence how we

conceive moral values and principles. Different metaphysical frameworks yield distinct ethical implications, shaping how individuals and societies understand right and wrong.

- ○ **Existentialism and Ethics**: Existentialist philosophers like **Jean-Paul Sartre** argue that individuals must create their own moral values in an indifferent universe. This perspective emphasises personal responsibility and authenticity, suggesting that ethical principles arise from individual choices rather than predetermined moral laws.

2. **Idealism and Ethics**

Idealism posits that reality is fundamentally mental or spiritual, suggesting that moral principles may arise from higher forms of consciousness or ideals. Philosophers such as **G.W.F. Hegel** propose that ethical development is tied to the unfolding of the Absolute Spirit, emphasising the importance of social and historical contexts in shaping moral values.

- ○ **The Role of Reason**: Idealism underscores the significance of reason in ethical decision-making, suggesting that moral principles must align with rational understanding. This perspective encourages individuals to engage with ethical dilemmas through reasoned analysis and reflection.

3. **Materialism and Ethics**
Materialist philosophies contend that the physical world is the primary reality, which influences ethical thought. From a materialist perspective, moral principles may emerge from human experiences and biological imperatives, emphasising the role of social and environmental factors in shaping moral behaviour.

- ○ **Consequentialism**: Materialist ethical frameworks often lead to consequentialist theories, which focus on the outcomes of actions in determining their moral worth. This approach evaluates ethical principles based on their consequences, advocating for actions that promote overall well-being.

4. **Metaphysical Implications of Moral Realism**

The acceptance of moral realism carries significant metaphysical implications. If moral truths exist independently, it suggests a realm of ethical reality that transcends individual beliefs. This view necessitates a deeper understanding of the relationship between moral principles and the nature of existence itself, prompting questions about how moral values interact with the fabric of reality.

The Nature of Good, Evil, and Balance

1. **Conceptualising Good and Evil**
   The definitions of good and evil vary
   across philosophical traditions, leading to
   differing ethical systems. Some traditions
   view good as that which promotes
   flourishing, happiness, or well-being,
   while evil is often defined as that which
   causes harm, suffering, or destruction.
   - **Utilitarianism**: This ethical
     framework, associated with
     philosophers like **Jeremy Bentham**
     and **John Stuart Mill**, defines good
     in terms of maximising happiness
     or utility. In this view, actions are
     evaluated based on their
     consequences for overall
     well-being, framing moral decisions
     in terms of their impact on the
     greatest number of people.

2. **The Duality of Existence**
   Many metaphysical systems acknowledge
   a duality between good and evil,
   suggesting that these forces are
   interconnected and necessary for

understanding existence. This dualistic perspective, found in various religious and philosophical traditions, posits that good and evil must coexist, creating a balance that is essential for the universe's functioning.

- ○ **Gnostic Thought**: Gnosticism presents a dualistic view, where the material world is seen as flawed or evil, in contrast to the spiritual realm of goodness. This perspective emphasises the importance of spiritual awakening and the pursuit of knowledge to transcend the material realm.

3. **The Role of Balance in Ethics**
   The idea of balance is central to many ethical philosophies, advocating for a harmonious approach to moral decision-making. The **Doctrine of the Mean**, articulated by **Aristotle**, suggests that virtue lies in finding a balanced response between extremes, promoting

moderation and reasoned judgement in ethical behaviour.

- ○ **Eastern Philosophies**: In Eastern traditions, such as **Taoism** and **Buddhism**, the concept of balance is pivotal. The Tao embodies the natural order of the universe, emphasising harmony between opposing forces. Similarly, Buddhist teachings advocate for the Middle Way, promoting a balanced approach to life that transcends extremes.

4. **The Interconnectedness of Good and Evil**

   Some philosophical perspectives argue that good and evil are not merely oppositional but are interdependent, each defining the other. This view suggests that the existence of evil enhances the understanding and appreciation of good, creating a dynamic interplay between these concepts.

○ **Theodicy**: Theodicy explores the relationship between God, good, and evil, attempting to justify the existence of evil in a world governed by a benevolent deity. Various theodicies propose that evil serves a purpose, such as fostering moral growth, testing faith, or enabling the development of virtues like compassion and resilience.

Meditation and Metaphysical Practices

Meditation is often regarded as a profound tool for self-discovery, personal growth, and spiritual development within the realm of metaphysics. Rooted in ancient traditions and philosophies, meditation serves as a bridge between the material and non-material worlds, offering practitioners insights into the nature of existence and their place within it. By engaging in meditation and other metaphysical practices, individuals can explore the depths of their consciousness, heighten their awareness, and tap into higher states of being.

Understanding Meditation as a Metaphysical Tool

1. Defining Meditation

Meditation is a practice that involves focused attention and a deep state of relaxation, allowing individuals to quiet the mind and explore the inner workings of consciousness. While various forms of meditation exist, all share the common goal of fostering a greater understanding of oneself and the universe. In a metaphysical context, meditation is seen as a means to transcend ordinary consciousness, facilitating deeper insights into the nature of reality and existence.

- **Historical Context**: The roots of meditation can be traced back to ancient spiritual traditions, including Hinduism, Buddhism, Taoism, and various indigenous practices. Each of these traditions recognizes the transformative power of meditation as a tool for self-realisation and spiritual awakening.

2. Metaphysical Implications of Meditation

Meditation is not merely a relaxation technique; it is a profound metaphysical practice that offers insights into the nature of reality. By entering altered states of consciousness, practitioners can explore fundamental questions about existence, awareness, and the interconnectedness of all things. This exploration can lead to experiences of unity with the universe, a sense of timelessness, and a deeper understanding of one's true nature.

- **Consciousness Exploration**: Through meditation, individuals can explore different layers of consciousness, moving beyond the superficial thoughts and distractions of daily life. This journey into deeper awareness opens the door to understanding the metaphysical aspects of being, such as the nature of the soul, the purpose of existence, and the interconnected web of life.

3. **The Role of Intention in Meditation**
   Intention is a critical element in meditation, influencing the depth of experience and the insights gained. By setting a clear intention—whether it be seeking clarity, healing, or spiritual growth—practitioners can align their meditative practice with their metaphysical goals. The power of intention helps to focus the mind, channel energy, and cultivate a receptive state for deeper understanding.

   o **Mindfulness and Presence**: A meditative practice rooted in intention fosters mindfulness, encouraging practitioners to remain present in the moment. This presence cultivates awareness of thoughts, feelings, and sensations, allowing individuals to observe their inner landscape without judgement. Through mindfulness, practitioners can gain insights into their habitual thought patterns and emotional responses, facilitating personal transformation.

Techniques for Heightening Awareness

1. **Mindfulness Meditation**
   Mindfulness meditation is a technique that emphasises being present in the moment without judgement. This practice encourages individuals to observe their thoughts, emotions, and sensations as they arise, fostering a sense of awareness and acceptance. By cultivating mindfulness, practitioners can deepen their connection to the present moment, allowing them to experience life with heightened clarity and understanding.
   - **Breath Awareness**: One of the most effective ways to enhance

mindfulness is through breath awareness. By focusing on the natural rhythm of the breath, individuals can anchor themselves in the present, quieting the mind and creating space for deeper insights.

2. **Transcendental Meditation (TM)**
Transcendental Meditation is a form of mantra meditation that involves the repetition of a specific sound or phrase. This technique allows the mind to settle into a state of deep relaxation, transcending ordinary thought processes. Through TM, practitioners can experience heightened states of awareness and access deeper layers of consciousness.

   ○ **The Power of Sound**: The use of sound and vibration in TM resonates with the metaphysical belief that all things are interconnected through energy and frequency. By attuning to specific sounds, individuals can align with higher states of consciousness and tap into the universal energy that permeates existence.

3. **Guided Visualisation**

   Guided visualisation is a technique that involves using imagery and imagination to create a desired mental state or experience. In a metaphysical context, guided visualisation can facilitate journeys into higher realms of consciousness, connecting practitioners with their inner wisdom, spirit guides, or universal truths.

   o **Creating Sacred Spaces**: During guided visualisation, individuals can create mental imagery of sacred spaces, such as gardens, temples, or celestial realms. These spaces serve as portals to deeper understanding and connection, allowing practitioners to explore metaphysical concepts and experiences.

4. **Breathwork and Energy Practices**

   Breathwork involves using conscious

breathing techniques to cultivate awareness and shift energy within the body. Various breathwork practices, such as **pranayama** in yoga or **holotropic breathwork**, can heighten awareness and facilitate altered states of consciousness.

- **Energy Flow and Awareness**: Breath is considered a vital force in many metaphysical traditions, believed to connect the physical and non-physical aspects of existence. By harnessing the power of breath, practitioners can enhance their energy flow, deepen their awareness, and facilitate experiences of unity with the universe.

Practical Exercises to Tap into Higher Consciousness

1. **Daily Mindfulness Practice**
   Set aside a few minutes each day to

practise mindfulness. Find a quiet space, sit comfortably, and focus on your breath. As thoughts arise, acknowledge them without judgement and gently return your focus to your breath. Over time, this practice will cultivate heightened awareness and an ability to remain present.

2. **Meditative Journaling**

   After a meditation session, take time to journal your experiences and insights. Write down any thoughts, feelings, or revelations that arose during your practice. This exercise not only reinforces your meditative experiences but also deepens your understanding of your inner world and metaphysical concepts.

3. **Guided Meditation Sessions**

   Explore guided meditations focused on specific metaphysical themes, such as connecting with your higher self, exploring past lives, or receiving messages from the universe. Many resources, including apps and online

platforms, offer guided meditations
tailored to various spiritual goals.

4. **Visualisation Exercises**
   Practice visualising a sacred space where
   you feel safe and connected. During
   meditation, allow yourself to explore this
   space, inviting in energies, symbols, or
   guides that resonate with your
   metaphysical journey. Observe any
   insights or sensations that arise during this
   exploration.

5. **Group Meditation**
   Join a meditation group or community that
   shares similar metaphysical interests.
   Group meditation can amplify energy and
   intention, facilitating deeper connections
   and shared experiences. Engaging with
   others can also provide support and
   encouragement on your spiritual journey.

6. **Nature Connection**
   Spend time in nature as a form of
   meditation. Observe the beauty and
   interconnectedness of the natural world,
   allowing yourself to feel the energy of the

environment. Nature can serve as a powerful reminder of the metaphysical principles of unity and harmony.

Quantum Physics and Metaphysics

Quantum physics, a revolutionary branch of science that explores the behaviour of matter and energy at the smallest scales, has significant implications for metaphysical thought. The intersection of quantum physics and metaphysics opens avenues for understanding existence, consciousness, and the nature of reality.

How Quantum Theory Relates to Metaphysical Thought

1. **Foundational Principles of Quantum Theory**

   Quantum theory emerged in the early 20th

century as scientists sought to explain phenomena that classical physics could not adequately address, such as the behaviour of subatomic particles. Key principles of quantum theory include wave-particle duality, superposition, and entanglement. These principles challenge traditional notions of reality, suggesting that particles can exist in multiple states simultaneously and that they are interconnected in ways that defy classical understanding.

- ○ **Wave-Particle Duality**: This principle posits that particles, such as electrons and photons, exhibit both wave-like and particle-like properties. This duality raises profound questions about the nature of reality: Are particles fundamental entities, or are they manifestations of underlying waves of potentiality? This idea resonates with metaphysical inquiries into the fundamental nature of existence.

- ○ **Superposition**: Superposition allows particles to exist in multiple states until measured or observed. This phenomenon echoes metaphysical concepts of potentiality and possibility, suggesting that reality is not fixed but rather a web of interconnected possibilities that collapse into a single outcome upon observation.
- ○ **Entanglement**: Quantum entanglement describes a phenomenon where particles become interconnected in such a way that the state of one particle instantaneously influences the state of another, regardless of the distance separating them. This concept challenges conventional notions of separateness and individuality, aligning with metaphysical perspectives on the interconnectedness of all existence.

2. Philosophical Implications

The principles of quantum theory challenge the deterministic worldview of classical physics, suggesting that reality is not a straightforward, objective phenomenon. Instead, it proposes a more complex and nuanced understanding of existence, resonating with various metaphysical traditions that emphasise the fluidity of reality and the role of consciousness in shaping experience.

- ○ **Metaphysical Realism vs. Idealism**: Quantum physics invites a reconsideration of metaphysical realism—the belief that reality exists independently of observers. Instead, some interpretations of quantum mechanics lean toward a form of idealism, wherein consciousness plays a crucial role in the manifestation of reality. This shift parallels philosophical debates about the nature of existence and the role of the observer in shaping reality.

- ○ **Non-Locality and Unity**: The phenomenon of entanglement suggests a non-locality in nature, where particles are interconnected across vast distances. This challenges the traditional understanding of space and separateness, aligning with metaphysical views that emphasise the unity of all existence and the interdependence of all things.

3. **Metaphysical Interpretations of Quantum Theory**
   Several interpretations of quantum mechanics highlight its metaphysical implications:
   - ○ **Copenhagen Interpretation**: This interpretation posits that particles do not have definite properties until they are observed. The act of observation collapses the wave function, determining the outcome

of a quantum event. This interpretation suggests a profound connection between consciousness and the fabric of reality, raising questions about the nature of perception and existence.

- ○ **Many-Worlds Interpretation**: This interpretation proposes that all possible outcomes of quantum events exist in parallel universes. Each decision or observation creates a branching of realities, leading to an infinite number of coexisting worlds. This perspective aligns with metaphysical ideas of potentiality and the multiverse, suggesting that reality is a vast tapestry of interconnected possibilities.

- ○ **Pilot-Wave Theory**: This deterministic interpretation introduces the concept of a guiding wave that influences particle behaviour. It posits that particles have definite trajectories, governed by a guiding wave. This interpretation emphasises the role of underlying laws in shaping reality, resonating with metaphysical inquiries into the nature of causality and order.

The Observer Effect and Reality

1. **Understanding the Observer Effect**
   The observer effect refers to the phenomenon where the act of observing a quantum system alters its state. This effect is famously illustrated in experiments such as the double-slit experiment, where particles exhibit wave-like behaviour when unobserved but behave like particles when measured.

   - **Measurement and Reality**: The observer effect raises fundamental questions about the nature of reality. If the act of observation influences

the outcome of a quantum event,
what does this imply about the
objective nature of existence? This
inquiry parallels metaphysical
discussions about the relationship
between perception and reality,
emphasising the role of
consciousness in shaping
experience.

2. **Implications for Understanding Reality**
The observer effect challenges the notion
of an objective reality independent of
perception. Instead, it suggests that reality
may be a co-creation between the observer
and the observed, highlighting the fluid
and dynamic nature of existence.

   ○ **Consciousness as a Creative
   Force**: In light of the observer
   effect, consciousness emerges as a
   potent force in shaping reality. This
   perspective aligns with
   metaphysical views that regard
   consciousness as fundamental to
   existence, where individual

awareness influences the
manifestation of the universe.

○ **Reality as Relational**: The
observer effect underscores the
relational nature of reality,
suggesting that our understanding
of existence is inherently
intertwined with perception and
experience. This aligns with
metaphysical perspectives that
emphasise the interconnectedness
of all beings and the significance of
subjective experience in
understanding the universe.

The Power of Thought and Intention

1. **Quantum Mechanics and Intention**
   The principles of quantum physics suggest
   that thought and intention may influence
   reality at a fundamental level. This idea
   resonates with various metaphysical

teachings that emphasise the power of consciousness in shaping experience.

- ○ **Thought as Energy**: In quantum theory, all matter is composed of energy. This perspective invites contemplation of the idea that thoughts and intentions, being forms of energy, may have the capacity to influence the fabric of reality. The notion that positive thoughts can lead to positive outcomes aligns with metaphysical beliefs about manifestation and the power of intention.

2. **Scientific Studies on Intention**
   Research in fields such as psychoneuroimmunology and quantum biology has begun to explore the potential connections between intention, consciousness, and physical reality. Studies have suggested that focused intention may influence biological processes, healing, and even the behaviour of quantum systems.

- ○ **Healing and Intention**: Various studies have demonstrated the potential for intention to impact healing outcomes. For instance, distant healing practices, where individuals direct their thoughts and intentions toward healing another person, have shown promising results in controlled studies. This phenomenon aligns with metaphysical principles of interconnectedness and the power of consciousness.

3. **Manifestation and Reality Creation**
   The concept of manifestation, popularised in metaphysical and spiritual communities, emphasises the ability to create one's reality through focused thought and intention. This aligns with quantum principles that suggest the observer plays a crucial role in shaping outcomes.

   - ○ **Visualisation Techniques**: Practitioners of manifestation often use visualisation techniques to align their thoughts and intentions with desired outcomes. By creating vivid mental images of their goals, individuals aim to influence their reality at both a personal and quantum level, echoing the idea that consciousness can shape existence.

4. **The Role of Belief**

Belief plays a significant role in shaping thought and intention. Metaphysical teachings emphasise that belief systems influence perception, reality, and the outcomes we experience. By cultivating positive beliefs and intentions, individuals can harness the power of consciousness to create desired changes in their lives.

- ○ **Transformative Potential**: The interplay between thought, belief, and intention aligns with the idea that individuals possess the power to transform their realities through conscious choices. This perspective empowers individuals to take an active role in shaping their experiences, resonating with metaphysical notions of self-determination and personal responsibility.

Alternate Dimensions and Multiverses

The concepts of alternate dimensions and multiverses have captivated the imagination of scientists, philosophers, and the general public alike. These ideas challenge our conventional understanding of reality and propose a more intricate and expansive framework for existence.

What are Alternate Dimensions?

1. **Defining Alternate Dimensions**
   Alternate dimensions refer to hypothetical realms of existence that coexist with our

familiar three-dimensional universe. In
these dimensions, the laws of physics may
differ, leading to distinct forms of matter,
energy, and even time. The concept
suggests that reality is not limited to the
physical universe we experience but is
part of a broader tapestry of existence.

- ○ **Dimensional Framework**: In
  physics, dimensions are typically
  defined as directions in which
  objects can move or be measured.
  While we experience three spatial
  dimensions (length, width, height)
  and one temporal dimension (time),
  theories of alternate dimensions
  propose additional dimensions that
  are either hidden or fundamentally
  different from our own. For
  example, string theory posits the
  existence of up to 11 dimensions,
  with additional dimensions curled
  up at scales so small that they
  remain undetectable by current
  technology.

- **Mathematical Models**:
  Mathematically, alternate
  dimensions can be represented in
  various ways. For instance, in string
  theory, the universe is depicted as a
  multi-dimensional object where
  different dimensions correspond to
  various physical properties. This
  mathematical framework allows
  physicists to explore the
  implications of additional
  dimensions on the nature of reality.

2. **Characteristics of Alternate Dimensions**
   The nature of alternate dimensions is
   largely speculative, but some
   characteristics can be proposed based on
   theoretical models:

   - **Different Physical Laws**: In
     alternate dimensions, the
     fundamental laws of physics may
     operate differently. This could
     result in unique forms of matter and
     energy, potentially leading to
     entirely different physical

phenomena. For example, in a dimension where gravity operates differently, celestial bodies might behave in unexpected ways.

- ○ **Nonlinear Time**: Time in alternate dimensions may not follow the linear progression experienced in our universe. Instead, time could be cyclical, multidimensional, or operate independently of space. This raises intriguing questions about causality and the nature of existence in such dimensions.

- ○ **Existence of Beings**: The possibility of sentient beings existing in alternate dimensions invites philosophical discussions about the nature of consciousness and identity. These beings might possess different forms of awareness or experience reality in ways fundamentally different from our own.

3. **Cultural and Philosophical Perspectives**
   Throughout history, the concept of
   alternate dimensions has appeared in
   various cultural and philosophical
   contexts. From ancient mythologies to
   modern science fiction, the idea of parallel
   realities has fueled creative explorations
   of existence.

   - **Mythology and Spirituality**: Many
     cultures have envisioned alternate
     realms or dimensions inhabited by
     gods, spirits, or otherworldly
     beings. These mythologies often
     serve to explain the mysteries of
     existence, death, and the nature of
     reality. For example, concepts of
     heaven, hell, and other spiritual
     dimensions reflect humanity's quest
     to understand what lies beyond the
     physical world.
   - **Philosophical Implications**:
     Philosophers have long pondered
     the nature of reality and existence.
     The notion of alternate dimensions

challenges traditional metaphysical views, prompting inquiries into the nature of truth, perception, and consciousness. If multiple dimensions exist, what does this imply about the nature of knowledge and reality itself?

Exploring the Theory of the Multiverse

1. **Defining the Multiverse**
The multiverse is a theoretical framework suggesting the existence of multiple, perhaps infinite, universes beyond our own. Each universe in the multiverse may have different physical constants, laws of nature, and even varying dimensions of time and space.
   - **Types of Multiverse Theories**: Various models of the multiverse exist, each based on different scientific theories. Some of the most prominent include:

■ **Cosmic Inflation Multiverse**: This model arises from the theory of cosmic inflation, which posits that the universe underwent rapid expansion in its early moments. In this framework, different regions of space could stop inflating at different times, leading to the formation of separate universes, each with its unique properties.

■ **Quantum Multiverse**: Rooted in quantum mechanics, this theory suggests that every time a quantum event occurs with multiple possible outcomes, a branching occurs, resulting in parallel universes for each possible outcome. This interpretation aligns with the

many-worlds interpretation
of quantum mechanics.

- **String Theory Multiverse**:
  String theory proposes a
  landscape of possible vacuum
  states, each representing a
  different universe with its
  unique physical laws. The
  vast number of solutions to
  string theory equations
  suggests a multiverse
  populated by diverse
  universes.

2. **Scientific Evidence and Speculation**
   While the multiverse remains a
   speculative concept, certain observations
   and theoretical frameworks provide
   indirect support for its plausibility.

   - **Cosmic Background Radiation**:
     Studies of cosmic background
     radiation have revealed anomalies
     that could suggest interactions
     between our universe and others.
     Some physicists theorise that these

anomalies might indicate the presence of neighbouring universes.

- ○ **Fine-Tuning of Constants**: The apparent fine-tuning of physical constants necessary for the existence of life has led some scientists to propose that a multiverse could account for this phenomenon. In a vast multiverse, it is conceivable that some universes are hospitable to life while others are not.

3. **Philosophical Implications of the Multiverse**

The multiverse theory raises profound philosophical questions about existence, identity, and the nature of reality.

- ○ **Identity and Individuality**: If multiple versions of ourselves exist in parallel universes, what does this mean for our sense of self? Philosophical discussions surrounding identity and the nature of consciousness become increasingly complex in a multiverse context.

- ○ **Reality and Truth**: The existence of multiple universes challenges the notion of an objective reality. If each universe operates under its laws, how do we define truth? This inquiry delves into the nature of perception and the role of consciousness in shaping our understanding of reality.

Parallel Universes and Their Implications

1. **Understanding Parallel Universes**
   Parallel universes are specific manifestations of the broader multiverse concept, representing alternate realities that exist alongside our own. Each parallel universe may differ in varying degrees, from minor variations to entirely distinct physical laws.

○ **Conceptual Framework**: In the context of quantum mechanics, parallel universes arise from the idea that every quantum event leads to a branching of realities. For instance, when faced with a decision, every possible outcome may create a new universe where that outcome is realised.

2. **Implications for Reality**

The existence of parallel universes poses significant questions about the nature of reality and our understanding of existence.

○ **Choices and Consequences**: The idea of parallel universes suggests that every choice we make creates a branch in reality, leading to different outcomes. This has implications for personal agency, free will, and the nature of destiny. Philosophically, it challenges our understanding of causality and the impact of individual choices.

○ **Experiencing Multiple Realities**:
If parallel universes exist,
individuals might be experiencing a
myriad of lives across different
realities. This invites inquiries into
the nature of consciousness and
whether awareness extends beyond
the confines of a single universe.

3. **Scientific and Philosophical
Considerations**
The implications of parallel universes
extend into scientific inquiry and
philosophical exploration.

○ **Scientific Validation**: While
empirical evidence for parallel
universes remains elusive, ongoing
research in quantum mechanics,
cosmology, and theoretical physics
continues to explore the feasibility
of such realities. Experiments
designed to test the principles of
quantum mechanics may yield
insights into the existence of
parallel universes.

○ **Philosophical Reflection**: The concept of parallel universes invites philosophical reflection on existence, identity, and the nature of reality. As we contemplate the possibility of multiple lives and realities, we are compelled to question the essence of self and the interconnectedness of all beings.

Mysticism and Spiritual Awakening

Mysticism and spiritual awakening are profound elements of metaphysical thought, offering insights into the nature of reality and the human experience. They delve into the relationships between consciousness, existence, and the universe, providing pathways for individuals to explore their inner selves and connect with the greater whole.

The Role of Mystics in Metaphysical Thought

1. **Defining Mysticism**

   Mysticism can be defined as the pursuit of communion with, or conscious awareness of, the ultimate reality, the divine, or transcendent truths. It transcends conventional religious practices, emphasising direct personal experiences of the divine or the universe. Mystics often seek to understand the nature of existence, consciousness, and the interplay between the individual and the cosmos.

2. **Historical Influence**

   Throughout history, mystics have played a crucial role in shaping metaphysical thought across various cultures and traditions. Their writings and experiences often challenge established doctrines, offering new perspectives on spirituality, existence, and the divine.

   - **Eastern Mysticism**: In traditions such as Hinduism, Buddhism, and Taoism, mystics have emphasised the importance of inner experience and meditation as pathways to enlightenment. For example, the teachings of mystics like Ramakrishna and the Zen masters highlight the direct experience of reality beyond intellectual understanding.

- ○ **Western Mysticism**: In the Western tradition, figures such as Meister Eckhart, St. John of the Cross, and Rumi explored the nature of the divine and human existence. Their works often emphasised the unity of all beings and the deep interconnectedness of life, challenging dualistic thinking and promoting a holistic understanding of reality.

3. **Contributions to Metaphysical Thought**
   Mystics contribute to metaphysical thought by offering experiential insights that challenge rational or dogmatic interpretations of reality. Their experiences often transcend language,

leading to profound insights about the
nature of existence and the divine.

- ○ **Unity Consciousness**: Many
  mystics articulate a vision of unity
  or interconnectedness among all
  beings. This perspective aligns with
  metaphysical concepts that
  emphasise the interrelation of all
  things and the idea that separation
  is an illusion. Mystics often
  describe states of consciousness
  where the self dissolves into the
  greater whole, reinforcing the
  metaphysical idea that individuality
  is part of a larger cosmic fabric.

- ○ **Transcendental Experiences**:
  Mystics frequently report
  transcendental experiences
  characterised by a profound sense
  of peace, love, and connection with
  the universe. These experiences
  often lead to radical shifts in
  perception and understanding,
  highlighting the potential for

personal transformation and
spiritual growth.

Stages of Spiritual Awakening

1. **Understanding Spiritual Awakening**
   Spiritual awakening refers to the process
   of becoming aware of one's true nature
   and the interconnectedness of all
   existence. It involves a profound shift in
   perception, leading individuals to seek a
   deeper understanding of themselves and
   the universe. While the journey of
   spiritual awakening is unique for each
   individual, it often unfolds through
   distinct stages.
2. **Stages of Awakening**
   Various models describe the stages of
   spiritual awakening, each reflecting
   different aspects of the transformative
   journey. A commonly referenced model
   includes the following stages:

○ **The Call**: The journey often begins with a sense of dissatisfaction or a longing for something more. Individuals may feel a disconnection from their surroundings or question the meaning of life. This initial call can manifest as a deep yearning for understanding, prompting individuals to explore spiritual paths.

○ **The Quest**: Following the call, individuals embark on a quest for knowledge, truth, and inner peace. This stage may involve exploring various spiritual practices, philosophies, or traditions. Individuals often seek guidance from mentors, books, or spiritual communities during this exploration.

○ **The Dark Night of the Soul**: Many individuals encounter challenges during their spiritual journey, often referred to as the "dark night of the soul." This phase can involve feelings of confusion, despair, or a sense of loss as old beliefs and attachments dissolve. This stage is crucial for growth, as it prompts individuals to confront their inner shadows and integrate their experiences.

○ **Awakening**: In this stage, individuals experience a significant shift in consciousness, leading to a profound understanding of their true nature. They may perceive reality more clearly, recognizing the interconnectedness of all beings and experiencing a sense of unity with the universe. This awakening often brings feelings of peace, joy, and unconditional love.

○ **Integration**: Following awakening, individuals enter a phase of integration, where they learn to embody their newfound understanding in daily life. This stage involves aligning actions with higher truths, nurturing compassion, and fostering connections with others. Individuals may engage in practices that support their spiritual growth, such as meditation, mindfulness, or community service.

○ **Ongoing Evolution**: Spiritual awakening is often viewed as an ongoing journey rather than a final destination. Individuals continue to evolve, deepening their understanding and experiencing further transformations. This ongoing process can lead to increased wisdom, compassion, and a commitment to personal and collective well-being.

3. **Personal Experiences**
Spiritual awakening can manifest in diverse ways, influenced by cultural, psychological, and experiential factors. Some individuals may experience sudden

shifts in awareness, while others may undergo gradual transformations. The uniqueness of each journey underscores the personal nature of spiritual awakening.

Pathways to Personal Enlightenment

1. **Exploring Various Pathways**
   The journey to personal enlightenment encompasses a myriad of pathways, each offering unique insights and experiences. While there is no one-size-fits-all approach, individuals can explore various methods to foster their spiritual growth and awakening.
2. **Meditation and Mindfulness**
   - **Meditation**: Practising meditation serves as a powerful tool for cultivating awareness and connecting with the present moment. Through focused attention and stillness, individuals can quiet the mind, explore inner landscapes,

and tap into deeper states of consciousness. Various forms of meditation, such as mindfulness meditation, loving-kindness meditation, and transcendental meditation, offer diverse approaches to exploring consciousness.

 o **Mindfulness**: Engaging in mindfulness practices helps individuals cultivate awareness in daily life, fostering a deeper connection with themselves and their surroundings. Mindfulness encourages individuals to be present, observe thoughts and emotions without judgement, and appreciate the richness of each moment. This heightened awareness can lead to transformative insights and a greater sense of peace.

3. **Study and Contemplation**

○ **Philosophical Exploration**:
Engaging with philosophical texts
and spiritual literature can deepen
understanding and stimulate
reflection. Exploring the writings of
mystics, philosophers, and spiritual
leaders can provide insights into
various metaphysical concepts and
foster a sense of connection with a
broader spiritual tradition.

○ **Journaling and Reflection**:
Keeping a journal can facilitate
personal exploration and
introspection. Individuals can
document their thoughts,
experiences, and insights, creating a
record of their spiritual journey.
Reflecting on one's experiences can
help clarify intentions and deepen
self-awareness.

4. **Community and Connection**

○ **Spiritual Communities**: Joining a
spiritual community or group can
provide support, encouragement,

and a sense of belonging. Engaging with like-minded individuals fosters shared exploration and collective growth, enabling individuals to learn from one another's experiences.

- ○ **Mentorship and Guidance**: Seeking guidance from experienced spiritual teachers or mentors can offer valuable insights and support on the spiritual journey. Mentors can provide direction, share their experiences, and help individuals navigate challenges along the way.

5. **Nature and Presence**
   - ○ **Connection with Nature**: Spending time in nature can facilitate a sense of connection to the universe and promote inner peace. Engaging with the natural world allows individuals to experience the beauty and interconnectedness of all living

things, fostering a deeper appreciation for existence.

○ **Present-Moment Awareness**: Cultivating a practice of present-moment awareness in daily activities encourages individuals to engage fully with life. By immersing themselves in each moment, individuals can deepen their connection to reality and experience the richness of existence.

Practical Applications of Metaphysics

Metaphysics, often regarded as an abstract field of philosophy concerned with the nature of reality and existence, has practical applications that can significantly impact daily life. By understanding and applying metaphysical principles, individuals can enhance their well-being, foster personal growth, and transform their perceptions of reality.

Using Metaphysical Principles in Daily Life

1. **Understanding the Interconnectedness of All Things**
   One of the fundamental principles of metaphysics is the understanding that all things are interconnected. Recognizing this interconnectedness can lead to more compassionate and mindful interactions with others and the environment. By adopting an awareness of how actions ripple through the larger web of existence, individuals can cultivate a sense of responsibility toward themselves, others, and the world around them.

2. **Embracing the Power of Thoughts and Beliefs**
   Metaphysics emphasises the influence of thoughts and beliefs on reality. The idea that our mental states shape our experiences can empower individuals to cultivate positive thinking and challenge limiting beliefs. By becoming aware of the narrative one tells oneself, individuals can replace negative self-talk with affirmations and empowering beliefs that

support personal growth and
transformation.

3. **Setting Intentions and Goals**
Utilising metaphysical principles involves
the practice of setting clear intentions for
various aspects of life. Intentions serve as
guiding beacons, aligning thoughts and
actions toward desired outcomes. By
articulating intentions—whether for
health, relationships, career, or personal
growth—individuals can focus their
energy and efforts on manifesting those
aspirations in their lives.

4. **Developing a Personal Philosophy**
Exploring metaphysical ideas can lead to
the development of a personal philosophy
that informs decision-making and life
choices. Individuals can draw on various
metaphysical perspectives to shape their
understanding of purpose, ethics, and
existence. This philosophy can serve as a
foundation for navigating challenges,
cultivating resilience, and fostering a
sense of meaning in life.

Mindfulness, Visualization, and Manifestation

1. **Mindfulness as a Metaphysical Practice**
   Mindfulness involves being fully present in the moment and aware of one's thoughts, feelings, and surroundings without judgement. This practice has deep roots in various spiritual traditions and is grounded in metaphysical principles that emphasise the importance of consciousness and awareness.

   - **Techniques for Mindfulness**: Individuals can cultivate mindfulness through meditation, breath awareness, and mindful observation of daily activities. Engaging in practices such as mindful eating, walking, or listening can enhance one's awareness of the present moment and deepen the understanding of self and reality.

○ **Benefits of Mindfulness**: Regular mindfulness practice can reduce stress, enhance emotional regulation, and improve overall well-being. By fostering a non-reactive awareness of thoughts and emotions, individuals can cultivate a sense of inner peace and clarity, aligning themselves more closely with their true essence.

2. **Visualisation as a Tool for Transformation**
Visualisation is a powerful technique often used in metaphysical practices to manifest desired outcomes. This involves creating mental images of specific goals or experiences, engaging the mind in a way that aligns with the intention to achieve those outcomes.

○ **Effective Visualisation Techniques**: Individuals can use guided imagery, vision boards, or focused visualisation meditations to enhance the effectiveness of their practice. By vividly imagining the desired outcome as if it has already been achieved, individuals can create a strong energetic alignment with their goals.

- ○ **The Science of Visualization**: Research has shown that visualisation can improve performance and outcomes in various fields, from athletics to personal development. By harnessing the power of the mind, individuals can influence their reality and manifest positive changes in their lives.

3. **Manifestation: Bringing Intentions into Reality**
Manifestation is the process of turning thoughts, intentions, and desires into reality. This practice draws on metaphysical principles, particularly the Law of Attraction, which posits that like attracts like and that individuals can draw

experiences into their lives through their thoughts and beliefs.

- ○ **The Process of Manifestation**: Successful manifestation involves several key steps:
  - ■ **Clarifying Intentions**: Clearly defining what one desires is crucial for effective manifestation. This clarity helps focus energy and intention on specific outcomes.
  - ■ **Aligning Thoughts and Emotions**: Ensuring that thoughts and emotions are aligned with the desired outcome is essential. Cultivating feelings of gratitude, joy, and belief in the possibility of achieving the intention can enhance the manifestation process.
  - ■ **Taking Inspired Action**: While thoughts and beliefs

are powerful, taking practical steps toward the desired outcome is equally important. Inspired action involves following intuitive nudges and opportunities that arise, aligning with the manifestation process.

- ○ **Overcoming Obstacles**: Manifestation may require overcoming limiting beliefs or fears that hinder progress. Engaging in self-reflection, therapy, or coaching can support individuals in addressing these barriers and aligning with their true desires.

Transforming Your Reality with Metaphysics

1. Shifting Perspectives

Metaphysical principles encourage individuals to shift their perspectives and challenge conventional beliefs about reality. By questioning assumptions and exploring alternative viewpoints, individuals can gain deeper insights into their experiences and transform their understanding of what is possible.

2. **Creating a Positive Environment**
   The environment we inhabit influences our thoughts and emotions. Metaphysics encourages individuals to create spaces that reflect their intentions and support their well-being. This can involve decluttering physical spaces, incorporating meaningful symbols or objects, and surrounding oneself with positive influences that resonate with personal values.

3. **Cultivating Gratitude and Acceptance**
   Practising gratitude and acceptance fosters a positive mindset that aligns with metaphysical principles. By appreciating what one has and accepting circumstances as they are, individuals can shift their focus away from negativity and scarcity. This perspective can attract more positive

experiences and foster a sense of abundance in life.

4. **Engaging in Metaphysical Practices**
Individuals can explore a variety of metaphysical practices to deepen their connection to the principles of existence and reality. Practices such as yoga, energy healing, crystal work, or ritual can enhance awareness and facilitate transformation. These practices allow individuals to tap into their inner resources, cultivate mindfulness, and align with their true selves.

5. **Living Authentically**
Embracing metaphysical principles encourages individuals to live authentically and express their true selves. By aligning thoughts, beliefs, and actions with their core values and desires, individuals can cultivate a life that reflects their essence. This authenticity fosters fulfilment and joy, contributing to a positive and meaningful existence.

Criticisms and Challenges in Metaphysics

Metaphysics, as a branch of philosophy, grapples with profound questions concerning existence, reality, and the nature of being. Despite its historical significance and depth, metaphysics faces numerous criticisms and challenges, ranging from misconceptions about its purpose to philosophical critiques questioning its validity.

Common Misconceptions About Metaphysics

1. **Metaphysics is Abstract and Irrelevant**
   One prevalent misconception is that metaphysics is overly abstract and disconnected from practical life. Critics often view metaphysical inquiries as esoteric musings that bear little relevance to everyday concerns. However, metaphysics seeks to explore foundational questions that underpin human experience, such as the nature of reality, the self, and existence itself. These inquiries can provide profound insights into personal and societal challenges, influencing ethical considerations, personal growth, and our understanding of the universe.

2. **Metaphysics is Just Speculation**
   Another common belief is that metaphysical thought is purely speculative, lacking empirical grounding. While metaphysics does engage with abstract concepts, it employs rigorous logical reasoning and philosophical argumentation. Furthermore, many

metaphysical claims intersect with scientific inquiries, prompting discussions that can lead to meaningful understanding. For example, debates about the nature of time, consciousness, and the universe often bridge metaphysical philosophy and scientific exploration, underscoring the relevance of metaphysical questions in a broader intellectual context.

3. **Metaphysics Conflicts with Science**
   Some argue that metaphysics inherently conflicts with scientific approaches, particularly when it deals with questions that science cannot empirically address. However, metaphysics and science can complement each other. Metaphysics explores the foundational principles and assumptions that underlie scientific theories, while science offers empirical data that can inform and refine metaphysical inquiries. The relationship between metaphysics and science is not one of opposition, but rather one of mutual influence and dialogue.

4. Metaphysics is Outdated

There is a perception that metaphysical questions are outdated, a relic of philosophical inquiry that has been supplanted by scientific advancements. Critics argue that as our understanding of the universe evolves through science, metaphysical questions become less relevant. Yet, metaphysical inquiry remains vital in addressing questions that science cannot definitively answer, such as the nature of consciousness, the existence of free will, and the ultimate purpose of life. As long as humans grapple with questions of existence and meaning, metaphysics will continue to hold relevance.

Philosophical Critiques of Metaphysical Thought

1. Empiricist Critique

Empiricism, a philosophical stance

asserting that knowledge derives from sensory experience, poses significant challenges to metaphysical claims. Prominent empiricists like David Hume criticised metaphysics for its reliance on speculative reasoning rather than empirical observation. Hume famously questioned the validity of causation, arguing that we can only observe sequences of events but cannot directly experience the causal relationships between them. This critique has led many to dismiss metaphysical claims as unfounded and lacking empirical support.

2. **Logical Positivism**

   The logical positivist movement, which emerged in the early 20th century, further scrutinised metaphysical inquiries. Proponents like A.J. Ayer contended that metaphysical statements are meaningless unless they can be verified through empirical observation or logical analysis. This perspective effectively disqualified much of metaphysical discourse as nonsensical. Critics of logical positivism, however, argue that this stance overlooks the rich philosophical insights that metaphysics can provide, highlighting the limitations of a purely empirical approach to understanding reality.

3. **Postmodern Critique**

   Postmodern philosophy introduces scepticism about the existence of objective truths, asserting that knowledge is shaped by cultural and social contexts. This perspective challenges traditional metaphysical assumptions about universal truths and fixed meanings. Postmodern thinkers, such as Jacques Derrida, argue that metaphysics tends to privilege certain narratives while marginalising others. This critique invites a more nuanced understanding of existence that acknowledges the fluidity of meaning and the role of perspective in shaping our understanding of reality.

4. **Existential and Phenomenological Critiques**

Existentialists and phenomenologists, like Jean-Paul Sartre and Martin Heidegger, challenge metaphysics by emphasising human experience and existence over abstract principles. They argue that traditional metaphysical frameworks often overlook the lived experiences of individuals and the existential conditions of being. Existentialism emphasises the importance of individual freedom and choice, while phenomenology focuses on the subjective nature of experience. These perspectives advocate for a more grounded approach to understanding reality, highlighting the significance of personal existence over abstract metaphysical constructs.

Addressing Scepticism and Misunderstandings

1. **Clarifying the Purpose of Metaphysics**
   To address scepticism, it is essential to

clarify the purpose and scope of metaphysics. Metaphysics seeks to explore fundamental questions about existence, reality, and the nature of being that are not easily confined to empirical investigation. By emphasising its relevance to personal and societal concerns, proponents can foster a deeper appreciation for the insights that metaphysical inquiry can provide.

2. **Promoting Dialogue Between Metaphysics and Science**
   Encouraging dialogue between metaphysical thought and scientific inquiry can help bridge perceived divides. By highlighting how metaphysical questions inform scientific theories and vice versa, advocates can demonstrate the complementary nature of both disciplines. This dialogue can lead to richer understandings of existence and reality, fostering collaboration between philosophers and scientists.

3. **Emphasising the Importance of Multiple Perspectives**
   Engaging with diverse philosophical perspectives, including empiricism, existentialism, and postmodernism, can enrich discussions around metaphysics. By acknowledging the value of different viewpoints, individuals can cultivate a more inclusive understanding of existence that embraces complexity and ambiguity. This openness to multiple perspectives can mitigate scepticism and foster appreciation for the nuances of metaphysical inquiry.

4. **Encouraging Personal Exploration**
   Ultimately, the study of metaphysics invites personal exploration and introspection. Individuals are encouraged to engage with metaphysical questions in ways that resonate with their experiences and beliefs. By fostering an environment that promotes inquiry, reflection, and open-mindedness, sceptics may come to

appreciate the transformative potential of metaphysical thought in their lives.

The Future of Metaphysics

Metaphysics, the philosophical inquiry into the nature of existence, reality, and the fundamental principles of being, continues to evolve in response to contemporary thought, scientific advancements, and cultural shifts. As we move further into the 21st century, the landscape of metaphysical inquiry is transforming, shaped by emerging trends, the interplay with modern science, and the exploration of new ideas. This section delves into the future of metaphysics, examining the trends that are reshaping the field, its role in contemporary science, and potential directions for further exploration.

Emerging Trends and Ideas

1. **Interdisciplinary Approaches**
   One of the most significant trends in the future of metaphysics is the increasing interdisciplinary collaboration among philosophers, scientists, and scholars from diverse fields. As complex questions arise at the intersections of metaphysics, physics, biology, and cognitive science, there is a growing recognition that addressing these questions requires collaborative inquiry. Philosophers are engaging with scientists to explore concepts such as consciousness, time, and the nature of reality, leading to richer, more nuanced understandings of these topics. This interdisciplinary dialogue fosters the development of new metaphysical frameworks that are informed by empirical research and scientific discovery.

2. **Quantum Metaphysics**
   The advent of quantum physics has

revolutionised our understanding of the universe and has significant implications for metaphysical thought. Concepts such as entanglement, superposition, and the observer effect challenge traditional notions of reality and causality. In the future, metaphysics may increasingly explore the implications of quantum theory, examining how these principles inform our understanding of existence, consciousness, and the interconnectedness of all things. This emerging field, often referred to as "quantum metaphysics," may lead to profound shifts in how we conceptualise reality itself.

3. **Technological Influence**
Rapid advancements in technology, particularly in artificial intelligence (AI) and virtual reality (VR), are creating new metaphysical questions and dilemmas. As AI systems become more sophisticated and capable of simulating human-like thought and behaviour, philosophers are prompted to consider the implications for

consciousness, identity, and what it means to "be." Additionally, VR technologies challenge our understanding of reality and perception, raising questions about the nature of experience and existence. The interplay between technology and metaphysical thought is likely to expand, offering new avenues for exploration and inquiry.

4. **Environmental Metaphysics**
   As global environmental challenges escalate, a growing movement within metaphysics is emerging that focuses on ecological and environmental concerns. This "environmental metaphysics" explores the relationship between humans and the natural world, examining concepts such as nature, sustainability, and our place within the ecosystem. Philosophers are increasingly questioning traditional dualistic views that separate humans from nature, advocating for a more holistic understanding of existence that recognizes the interconnectedness of all living beings.

This trend reflects a broader cultural shift toward environmental consciousness and sustainability.

5. **Metaphysics of Consciousness**

   The nature of consciousness remains one of the most profound and debated topics in metaphysical inquiry. Emerging research in neuroscience and psychology is shedding light on the complexities of consciousness, prompting philosophers to reevaluate traditional metaphysical frameworks. Future metaphysical exploration may delve deeper into the relationship between consciousness, self-awareness, and reality, examining how our understanding of consciousness influences our conception of existence. This evolving discourse may bridge the gap between philosophical inquiry and scientific investigation, leading to a more comprehensive understanding of the nature of mind.

The Role of Metaphysics in Modern Science

1. **Philosophy of Science**
   Metaphysics plays a crucial role in the philosophy of science, offering foundational insights that shape scientific theories and methodologies. Many scientific inquiries rest on metaphysical assumptions about the nature of reality, causation, and the structure of the universe. As scientific paradigms shift, so too do the metaphysical frameworks that underpin them. Future developments in physics, biology, and other scientific fields will likely prompt ongoing philosophical reflection on the metaphysical implications of new discoveries.

2. **Theoretical Physics and Metaphysics**
   Theoretical physics, particularly in areas such as quantum mechanics and cosmology, is increasingly engaging with metaphysical questions. Concepts like the nature of time, space, and the multiverse

are not only scientific inquiries but also deeply metaphysical ones. As physicists propose new models to explain the universe, metaphysicians will need to engage with these ideas, examining their implications for our understanding of existence and reality. This collaboration between physics and metaphysics can lead to a more comprehensive worldview that integrates empirical findings with philosophical inquiry.

3. **Ethics and Responsibility**
   As technology and scientific advancements progress, metaphysical thought will play an important role in addressing ethical concerns. Questions about the implications of AI, genetic engineering, and environmental sustainability require philosophical reflection on the nature of agency, responsibility, and the ethical treatment of living beings. Metaphysics can provide the conceptual tools needed to navigate these complex issues, helping society

formulate ethical frameworks that align with our evolving understanding of reality and existence.

Where Metaphysics is Headed Next

1. **New Metaphysical Frameworks**
   As metaphysical inquiry continues to evolve, it is likely that new frameworks will emerge that integrate insights from various disciplines. These frameworks may seek to reconcile traditional metaphysical concepts with contemporary scientific understanding, offering holistic perspectives that account for both empirical evidence and philosophical inquiry. The future of metaphysics may involve the development of pluralistic models that embrace diverse viewpoints and encourage collaboration across fields.

2. **Revisiting Classic Questions**
   While contemporary issues will shape the future of metaphysics, it is essential to

revisit classic questions that have long intrigued philosophers. Timeless inquiries into the nature of existence, the self, and the meaning of life will continue to resonate as individuals seek answers in an increasingly complex world. Future metaphysical explorations may reinvigorate these questions, drawing connections between historical thought and modern concerns.

3. **Global Perspectives**
   The future of metaphysics may also be shaped by an increasing emphasis on global perspectives. Philosophical traditions from various cultures, including Eastern philosophies such as Buddhism and Taoism, offer alternative metaphysical frameworks that challenge Western-centric views. Engaging with diverse philosophical traditions can enrich metaphysical inquiry, leading to a more comprehensive understanding of existence that transcends cultural boundaries.

4. **Public Engagement and Accessibility**

   As interest in metaphysical thought grows, there is an opportunity for greater public engagement and accessibility. Initiatives that promote metaphysical inquiry in educational settings, online platforms, and community discussions can foster a broader understanding of the subject. By making metaphysics more accessible, individuals from all backgrounds can explore its relevance to their lives, promoting a culture of inquiry and philosophical exploration.

Conclusion: Embracing the Mysteries of Existence

As we conclude our exploration of metaphysics, we find ourselves at the intersection of profound questions and the mysteries that define our existence. This journey through the intricacies of reality, being, and consciousness invites us to reflect not only on the concepts we have encountered but also on our place within the vast tapestry of the universe. Metaphysics offers a framework through which we can examine the deeper meanings of life, challenge our assumptions, and engage with the unknown.

Recap of Key Metaphysical Concepts

1. **The Nature of Reality and Existence**
   At the heart of metaphysical inquiry is the exploration of what it means to exist. We have examined the distinction between material and non-material worlds, the interplay between appearance and reality, and the various perspectives that inform our understanding of existence. This foundational concept invites us to question the nature of the world around us and the assumptions we hold about what is "real."

2. **The Nature of Being**
   Understanding what it means to "be" is central to metaphysics. We delved into ontology, the study of being, and the relationship between existence and consciousness. This exploration reveals

the complexity of self-awareness and the nature of identity, prompting us to consider our own experiences of being in the world.

3. **Time and Space**

   The concepts of time and space challenge our perceptions of reality. We discussed how time may be perceived differently, either as a linear progression or a more fluid experience, and examined the nature of space as it relates to existence. This inquiry encourages us to think critically about how we navigate our lives within these dimensions.

4. **Mind and Matter**

   The relationship between mind and body, and the philosophical debates surrounding dualism, materialism, and idealism, highlight the intricacies of consciousness and perception. Understanding how our thoughts and experiences shape our reality is a crucial aspect of metaphysical thought, guiding us toward greater self-awareness and insight.

5. **Causality and Free Will**

   The discussions surrounding causality and free will prompt us to consider the nature of choice and destiny. We examined how metaphysics addresses the complexities of human agency and the implications of determinism versus free will, revealing the profound impact our choices have on our lives.

6. **The Concept of the Soul**

   The exploration of the soul and its significance across various philosophical traditions invites us to reflect on the nature of life, death, and the possibility of an afterlife. This discussion encourages a deeper understanding of our existence and the potential for continued consciousness beyond physical life.

7. **Ethics and Morality**

   Metaphysics also engages with ethical questions, exploring the origins of right and wrong and the nature of good and evil. This inquiry fosters a greater understanding of our moral responsibilities and the interconnectedness of all beings within the universe.

8. **Practical Applications of Metaphysics**
Finally, we explored how metaphysical principles can be applied to daily life. Practices such as mindfulness, visualisation, and meditation provide tools for transformation, enabling us to harness metaphysical insights to shape our reality and deepen our understanding of existence.

How to Continue Your Metaphysical Journey

1. **Engage with Philosophical Texts**
One of the most rewarding ways to continue your metaphysical journey is by engaging with philosophical texts and literature. Delving into the works of influential metaphysicians—such as Aristotle, Descartes, Kant, and contemporary philosophers—will deepen your understanding of the foundational

concepts of metaphysics. Look for texts that challenge your perspective and invite you to think critically about existence and reality.

2. **Explore Different Philosophical Traditions**

   The diversity of philosophical thought across cultures offers rich insights into metaphysical questions. Investigate Eastern philosophies, such as Buddhism and Taoism, as well as indigenous philosophies that emphasise interconnectedness and the natural world. Engaging with these perspectives can provide a broader understanding of metaphysical concepts and how they manifest in various cultural contexts.

3. **Practice Meditation and Mindfulness**

   Incorporating meditation and mindfulness practices into your daily routine can enhance your awareness of existence and consciousness. These practices allow for introspection and self-discovery, helping you tap into deeper levels of understanding and connection to the universe. As you cultivate mindfulness, you may gain insights into the nature of reality and your place within it.

4. **Engage in Dialogues and Discussions**
Participating in discussions with others who are interested in metaphysical thought can provide new perspectives and insights. Join philosophy groups, attend workshops, or engage in online forums where metaphysical questions are explored. Sharing ideas and experiences with others can enrich your understanding and inspire new avenues of inquiry.

5. **Reflect on Personal Experiences**
Take time to reflect on your personal experiences and how they relate to metaphysical concepts. Consider moments of insight, intuition, or profound understanding that have shaped your view of existence. Keeping a journal can help you document these reflections and track your evolving understanding of metaphysical questions.

6. **Stay Curious and Open-Minded**

   The journey of metaphysical exploration
   is one of ongoing inquiry. Approach the
   mysteries of existence with curiosity and
   an open mind, recognizing that not all
   questions have definitive answers.
   Embrace the complexities and
   uncertainties of metaphysical thought,
   allowing your understanding to evolve as
   you encounter new ideas and experiences.

Glossary of Terms

1. **Metaphysics**

   Metaphysics is the branch of philosophy that studies the fundamental nature of reality, existence, and the relationship between mind and matter. It seeks to understand concepts such as being, time, space, causality, and the nature of the universe beyond empirical observation.

2. **Ontology**

   Ontology is the study of being and existence. It involves exploring the categories of being and the relationships between them. Questions in ontology include what entities exist and how they can be classified, such as distinguishing between physical objects, abstract concepts, and modes of existence.

3. **Epistemology**

   Epistemology is the branch of philosophy concerned with knowledge—its nature, sources, limits, and validity. It examines questions such as: What is knowledge? How do we acquire it? How do we know what we know? It also investigates the distinction between belief and knowledge.

4. **Reality**

   Reality refers to the state of things as they exist, independent of perception or belief. In metaphysics, discussions about reality often explore the nature of existence and what constitutes "real" versus "illusion" or "appearance."

5. **Existence**

   Existence is the state of being real or having actual being. Metaphysicians explore various forms of existence, including physical, abstract, and potential existence, and how these categories interact.

6. **Materialism**

   Materialism is a philosophical viewpoint asserting that only material (physical) things truly exist, and everything, including consciousness, can be explained in terms of physical processes. It contrasts with idealism, which posits that reality is fundamentally mental or immaterial.

7. **Idealism**

   Idealism is a philosophical theory that asserts that reality is primarily mental or immaterial. According to idealists, objects do not exist independently of the minds that perceive them, suggesting that the mind shapes reality.

8. **Dualism**

   Dualism is the belief that there are two fundamental types of substance or reality: physical (matter) and non-physical (mind or spirit). This view is often associated with Descartes, who argued for the distinction between the mind (thinking substance) and the body (extended substance).

9. **Causality**

    Causality refers to the relationship
    between cause and effect, where one event
    (the cause) leads to another event (the
    effect). In metaphysics, causality raises
    questions about determinism (the idea that
    every event is caused by previous events)
    and free will.

10. **Free Will**

    Free will is the ability to make choices
    unconstrained by external factors or fate.
    It is a central topic in discussions of moral
    responsibility, ethics, and the nature of
    human agency in a potentially
    deterministic universe.

11. **The Soul**

    The soul is often regarded as the
    immaterial essence of a person,
    representing their identity, consciousness,
    or spiritual aspect. Various philosophical
    and religious traditions provide different
    interpretations of the soul, its nature, and
    its fate after death.

12. Consciousness

Consciousness is the state of being aware of and able to think about one's own existence, thoughts, and surroundings. It is a central topic in metaphysical discussions regarding the relationship between mind and body.

13. Space

Space refers to the boundless three-dimensional extent in which objects and events occur and have relative position and direction. Metaphysical discussions about space often explore whether it is a tangible entity or merely a conceptual framework for understanding the relationships between objects.

14. Time

Time is the measurable period during which events occur, often perceived as a continuous progression from the past, through the present, and into the future. Philosophical inquiries into the nature of time explore whether it is absolute, relative, or an illusion.

15. **Phenomenology**

    Phenomenology is a philosophical approach that emphasises the study of conscious experience and the objects of direct experience. It seeks to understand how individuals perceive and interpret the world around them.

16. **Mysticism**

    Mysticism refers to spiritual experiences or practices that seek to achieve a direct, personal union with the divine or ultimate reality. Mystics often report transcendent experiences that challenge conventional notions of reality and existence.

17. **Pantheism**

    Pantheism is the belief that the divine pervades all aspects of the universe and that God is synonymous with nature and the universe itself. This view emphasises the interconnectedness of all things and the divine presence in the natural world.

18. **Determinism**

    Determinism is the philosophical view that all events, including human actions, are determined by preceding events and conditions, often governed by natural laws. This concept raises questions about the existence of free will and moral responsibility.

19. **Karma**

Karma is a concept found in various spiritual traditions that refers to the law of moral cause and effect, where an individual's actions (good or bad) influence their future experiences and circumstances. It is often associated with ideas of rebirth and spiritual progression.

20. **Multiverse**

The multiverse is a theoretical concept suggesting that there are multiple, perhaps infinite, universes existing alongside our own. Each universe may have different laws of physics, constants, or histories, leading to varied realities and experiences.

21. **Quantum Physics**

Quantum physics is a fundamental theory in physics describing the physical properties of nature at the scale of atoms and subatomic particles. It has significant implications for metaphysics, particularly regarding the nature of reality, observation, and the interconnectedness of all things.

22. **Vibrational Frequencies**

Vibrational frequencies refer to the rates at which particles or waves oscillate. In metaphysical thought, this concept is often associated with the idea that all matter and energy vibrate at specific frequencies, influencing their interactions and the nature of reality.

23. **Interconnectedness**

Interconnectedness is the concept that all things in the universe are related and influence one another. This idea is fundamental in many metaphysical systems, emphasising that actions, thoughts, and energies impact the broader whole.

24. **Meditation**

Meditation is a practice that involves focusing the mind and eliminating distractions to achieve a heightened state of awareness and mental clarity. In metaphysics, meditation is often viewed as a tool for exploring consciousness and achieving deeper insights into the nature of reality.

25. **Spiritual Awakening**

Spiritual awakening is a transformative process in which an individual gains insight into their true nature and the nature of reality. This process often involves a shift in consciousness, leading to greater awareness, purpose, and connection to the universe.

26. **Mystical Experience**

A mystical experience is a profound, often ineffable experience characterised by a sense of unity with the universe, transcendence of time and space, and a deep feeling of peace and love. Such experiences challenge conventional understandings of reality and can lead to significant personal transformation.

Philosophy Beyond the Physical

Philosophy Beyond the Physical

www.ingramcontent.com/pod-product-compliance
Lightning Source LLC
Chambersburg PA
CBHW052312220526
45472CB00001B/83